Lecture Notes in Mathematics 2026

Editors:
J.-M. Morel, Cachan
B. Teissier, Paris

Subseries:
École d'Été de Probabilités de Saint-Flour

For further volumes:
http://www.springer.com/series/304

Saint-Flour Probability Summer School

The Saint-Flour volumes are reflections of the courses given at the Saint-Flour Probability Summer School. Founded in 1971, this school is organised every year by the Laboratoire de Mathématiques (CNRS and Université Blaise Pascal, Clermont-Ferrand, France). It is intended for PhD students, teachers and researchers who are interested in probability theory, statistics, and in their applications.

The duration of each school is 13 days (it was 17 days up to 2005), and up to 70 participants can attend it. The aim is to provide, in three high-level courses, a comprehensive study of some fields in probability theory or Statistics. The lecturers are chosen by an international scientific board. The participants themselves also have the opportunity to give short lectures about their research work.

Participants are lodged and work in the same building, a former seminary built in the 18th century in the city of Saint-Flour, at an altitude of 900 m. The pleasant surroundings facilitate scientific discussion and exchange.

The Saint-Flour Probability Summer School is supported by:

– Université Blaise Pascal
– Centre National de la Recherche Scientifique (C.N.R.S.)
– Ministère délégué à l'Enseignement supérieur et à la Recherche

For more information, see back pages of the book and
http://math.univ-bpclermont.fr/stflour/

Jean Picard
Summer School Chairman
Laboratoire de Mathématiques
Université Blaise Pascal
63177 Aubiére Cedex
France

Yves Le Jan

Markov Paths, Loops and Fields

École d'Été de Probabilités
de Saint-Flour XXXVIII-2008

 Springer

Yves Le Jan
Université Paris-Sud
Département de Mathématiques
Bât.425
91405 Orsay Cedex
France
yves.lejan@math.upsud.fr

ISBN 978-3-642-21215-4   e-ISBN 978-3-642-21216-1
DOI 10.1007/978-3-642-21216-1
Springer Heidelberg Dordrecht London New York

Lecture Notes in Mathematics ISSN print edition: 0075-8434
                              ISSN electronic edition: 1617-9692

Library of Congress Control Number: 2011932434

Mathematics Subject Classification (2011): Primary 60J27, 60K35; Secondary 60J45

Cover design: deblik, Berlin

Printed on acid-free paper

Springer is part of Springer Science+Business Media (www.springer.com)

Preface

The purpose of these notes is to explore some simple relations between Markovian path and loop measures, the Poissonian ensembles of loops they determine, their occupation fields, uniform spanning trees, determinants, and Gaussian Markov fields such as the free field. These relations are first studied in complete generality in the finite discrete setting, then partly generalized to specific examples in infinite and continuous spaces.

These notes contain the results published in [27] where the main emphasis was put on the study of occupation fields defined by Poissonian ensembles of Markov loops. These were defined in [18] for planar Brownian motion in relation with SLE processes and in [19] for simple random walks. They appeared informally already in [52]. For half integral values $\frac{k}{2}$ of the intensity parameter α, these occupation fields can be identified with the sum of squares of k copies of the associated free field (i.e. the Gaussian field whose covariance is given by the Green function). This is related to Dynkin's isomorphism (cf. [6, 23, 33]).

As in [27], we first present the theory in the elementary framework of symmetric Markov chains on a finite space. After some generalities on graphs and symmetric Markov chains, we study the σ-finite loop measure associated to a field of conductances. Then we study geodesic loops with an exposition of results of independent interest, such as the calculation of Ihara's zeta function. After that, we turn our attention to the Poisson process of loops and its occupation field, proving also several other interesting results such as the relation between loop ensembles and spanning trees given by Wilson algorithm and the reflection positivity property. Spanning trees are related to the fermionic Fock space as Markovian loop ensembles are related to the bosonic Fock space, represented by the free field. We also study the decompositions of the loop ensemble induced by the excursions into the complement of any given set.

Then we show that some results can be extended to more general Markov processes defined on continuous spaces. There are no essential difficulties for the occupation field when points are not polar but other cases are more problematic. As for the square of the free field, cases for which the Green function is Hilbert Schmidt such as those corresponding to two and three dimensional Brownian motion can be dealt with through appropriate renormalization.

We show that the renormalized powers of the occupation field (i.e. the self intersection local times of the loop ensemble) converge in the case of the two dimensional Brownian motion and that they can be identified with higher even Wick powers of the free field when α is a half integer.

At first, we suggest the reader could omit a few sections which are not essential for the understanding of the main results. These are essentially some of the generalities on graphs, results about wreath products, infinite discrete graphs, boundaries, zeta functions, geodesics and geodesic loops. The section on reflexion positivity, and, to a lesser extent, the one on decompositions are not central. The last section on continuous spaces is not written in full detail and may seem difficult to the least experienced readers.

These notes include those of the lecture I gave in St Flour in July 2008 with some additional material. I choose this opportunity to express my thanks to Jean Picard, to the audience and to the readers of the preliminary versions whose suggestions were very useful, in particular to Juergen Angst, Cedric Bordenave, Cedric Boutiller, Jinshan Chang, Antoine Dahlqvist, Thomas Duquesne, Michel Emery, Jacques Franchi, Hatem Hajri, Liza Jones, Adrien Kassel, Rick Kenyon, Sophie Lemaire, Thierry Levy, Titus Lupu, Gregorio Moreno, Jay Rosen (who pointed out a mistake in the expression of renormalization polynomials), Bruno Shapira, Alain Sznitman, Vincent Vigon, Lorenzo Zambotti and Jean Claude Zambrini.

Contents

Chapter 1
Symmetric Markov Processes on Finite Spaces

Notations: functions and measures on finite (or countable) spaces are often denoted as vectors and covectors, i.e. with upper and lower indices, respectively.

The multiplication operator defined by a function f acting on functions or on measures is in general simply denoted by f, but sometimes, to avoid confusion, it will be denoted by M_f. The function obtained as the density of a measure μ with respect to some other measure ν is simply denoted $\frac{\mu}{\nu}$.

1.1 Graphs

Our basic object will be a finite space X and a set of non negative *conductances* $C_{x,y} = C_{y,x}$, indexed by pairs of distinct points of X. This situation allows to define a kind of discrete topology and geometry. In this first section, we will briefly study the topological aspects.

We say that $\{x, y\}$, for $x \neq y$ belonging to X, is a link or an edge iff $C_{x,y} > 0$. An oriented edge (x, y) is defined by the choice of an ordering in an edge. We set $-(x, y) = (y, x)$ and if $e = (x, y)$, we denote it also (e^-, e^+). The degree d_x of a vertex x is by definition the number of edges incident at x.

The points of X together with the set of non oriented edges E define a graph (X, E). *We assume it is connected.* The set of oriented edges is denoted E^o. It will always be viewed as a subset of X^2, without reference to any imbedding.

The associated line graph is the oriented graph defined by E^o as set of vertices and in which oriented edges are pairs (e_1, e_2) such that $e_1^+ = e_2^-$. The mapping $e \to -e$ is an involution of the line graph.

An important example is the case in which conductances are equal to zero or one. Then the conductance matrix is the adjacency matrix of the graph:
$$C_{x,y} = 1_{\{x,y\} \in E}$$
A complete graph is defined by all conductances equal to one.

Y. Le Jan, *Markov Paths, Loops and Fields*, Lecture Notes in Mathematics 2026, DOI 10.1007/978-3-642-21216-1_1, © Springer-Verlag Berlin Heidelberg 2011

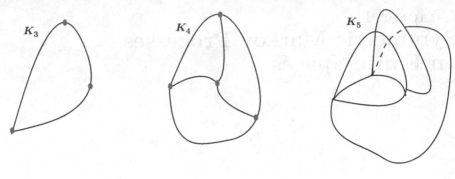

Fig. 1.1 Complete graphs

The complete graph with n vertices is denoted K_n. The complete graph K_4 is the graph defined by the tetrahedron. K_5 is not planar (i.e. cannot be imbedded in a plane), but K_4 is (Fig. 1.1).

A finite discrete path on X, say $(x_0, x_1, ..., x_n)$ is called a (discrete) *geodesic arc* iff $\{x_i, x_{i+1}\} \in E$ (path segment on the graph) and $x_{i-1} \neq x_{i+1}$ (without backtracking). Geodesic arcs starting at x_0 form a *marked tree* $\mathfrak{T}_{x_0}$ rooted in x_0 (the marks belong to X: they are the endpoints of the geodesic arcs). Oriented edges of $\mathfrak{T}_{x_0}$ are defined by pairs of geodesic arcs of the form: $((x_0, x_1, ..., x_n), (x_0, x_1, ..., x_n, x_{n+1}))$ (the orientation is defined in reference to the root). $\mathfrak{T}_{x_0}$ is a *universal cover* of X [34].

A (discrete) loop based at $x_0 \in X$ is by definition a path $\xi = (\xi_1, ..., \xi_{p(\xi)})$, with $\xi_1 = x_0$, and $\{\xi_i, \xi_{i+1}\} \in E$, for all $1 \leq i \leq p$ with the convention $\xi_{p+1} = \xi_1$. On the space $\mathfrak{L}_{x_0}$ of discrete loops based at some point x_0, we can define an operation of concatenation, which provides a monoid structure, i.e. is associative with a neutral element (the empty loop). The concatenation of two closed geodesics (i.e. geodesic loops) based at x_0 is not directly a closed geodesic. It can involve backtracking "in the middle" but then after cancellation of the two inverse subarcs, we get a closed geodesic, possibly empty if the two closed geodesics are identical up to reverse order. With this operation, *closed geodesics based at x_0 define a group* Γ_{x_0}. The structure of Γ_{x_0} does not depend on the base point and defines the *fundamental group* Γ of the graph (as the graph is connected: see for example [34]). Indeed, any geodesic arc γ_1 from x_0 to another point y_0 of X defines an isomorphism between Γ_{x_0} and Γ_{y_0}. It associates to a closed geodesic γ based in x_0 the closed geodesic $[\gamma_1]^{-1}\gamma\gamma_1$ (here $[\gamma_1]^{-1}$ denotes the backward arc). In the case where $x_0 = y_0$, it is an interior isomorphism (conjugation by γ_1).

There is a natural left action of Γ_{x_0} on $\mathfrak{T}_{x_0}$. It can be interpreted as a change of root in the tree (with the same mark). Besides, any geodesic arc between x_0 and another point y_0 of X defines an isomorphism between $\mathfrak{T}_{x_0}$ and $\mathfrak{T}_{y_0}$ (change of root, with different marks).

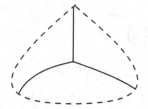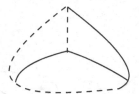

Fig. 1.2 Two spanning trees of K_4

We have just seen that the universal covering of the finite graph (X, E) at x_0 is a tree $\mathfrak{T}_{x_0}$ projecting on X. The fiber at x_0 is Γ_{x_0}. The groups $\Gamma_{x_0}, x_0 \in X$ are conjugated in a non canonical way. Note that $X = \Gamma_{x_0} \backslash \mathfrak{T}_{x_0}$ (here the use of the quotient on the left corresponds to the left action).

Example 1. Among graphs, the simplest ones are r-regular graphs, in which each point has r neighbours. A universal covering of any r-regular graph is isomorphic to the r-regular tree $\mathfrak{T}^{(r)}$.

Example 2. Cayley graphs: a finite group with a set of generators $S = \{g_1, ..g_k\}$ such that $S \cap S^{-1}$ is empty defines an oriented $2k$-regular graph.

A *spanning tree* T is by definition a subgraph of (X, E) which is a tree and covers all points in X. It has necessarily $|X| - 1$ edges, see for example two spanning trees of K_4 (Fig. 1.2).

The inverse images of a spanning tree by the canonical projection from a universal cover $\mathfrak{T}_{x_0}$ onto X form a tesselation on $\mathfrak{T}_{x_0}$, i.e. a partition of $\mathfrak{T}_{x_0}$ in identical subtrees, which are fundamental domains for the action of Γ_{x_0}. Conversely, a section of the canonical projection from the universal cover defines a spanning tree.

Fixing a spanning tree determines a unique geodesic between two points of X. Therefore, it determines the conjugation isomorphisms between the various groups Γ_{x_0} and the isomorphisms between the universal covers $\mathfrak{T}_{x_0}$.

Remark 1. Equivalently, we could have started with an infinite tree $\mathfrak{T}$ and a group Γ of isomorphisms of this tree such that the quotient graph $\Gamma \backslash \mathfrak{T}$ is finite.

The fundamental group Γ is a free group with $|E| - |X| + 1 = r$ generators. To construct a set of generators, one considers a spanning tree T of the graph, and choose an orientation on each of the r remaining links. This defines r oriented cycles on the graph and a system of r generators for the fundamental group. (See [34] or Serres [41] in a more general context).

Example 3. Consider K_3 and K_4.

Here is a picture of the universal covering of K_4, and of the action of the fundamental group with the tesselation defined by a spanning tree (Fig. 1.3).

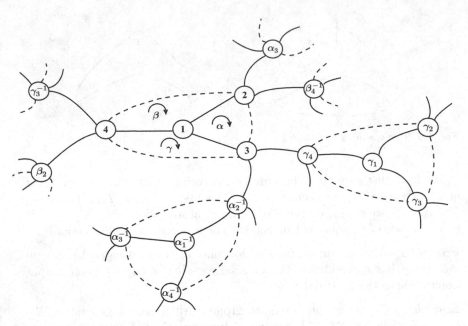

Fig. 1.3 Universal cover and tesselation of K_4

There are various non-ramified coverings, intermediate between (X, E) and the universal covering. Non ramified means that locally, the covering space is identical to the graph (same incident edges). Then each oriented path segment on X can be lifted to the covering in a unique way, given a lift of its starting point.

Each non ramified covering is (up to an isomorphism) associated with a subgroup H of Γ, defined up to conjugation. More precisely, given a non ramified covering $\widetilde{X}$, a point x_0 of X and a point $\widetilde{x}_0$ in the fiber above x_0, the closed geodesics based at x_0 whose lift to the covering starting at $\widetilde{x}_0$ are closed form a subgroup $H_{\widetilde{x}_0}$ of Γ_{x_0}, canonicaly isomorphic to the fundamental group of $\widetilde{X}$ represented by closed geodesics based at $\widetilde{x}_0$. If we consider a different point $\widetilde{y}_0$, any geodesic path segment $\widetilde{\gamma}_1$ between $\widetilde{x}_0$ and $\widetilde{y}_0$ defines an isomorphism between Γ_{x_0} and Γ_{y_0} which exchanges $H_{\widetilde{x}_0}$ and $H_{\widetilde{y}_0}$. Denoting γ_1 the projection of $\widetilde{\gamma}_1$ on X, it associates to a closed geodesic γ based in x_0 whose lift to the covering is closed the closed geodesic $[\gamma_1]^{-1}\gamma\gamma_1$ whose lift to the covering is also closed.

Example 4. By central symmetry, the cube is a twofold covering of the tetrahedron associated with the group $\mathbb{Z}/2\mathbb{Z}$.

Conversely, if H is a subgroup of Γ_{x_0}, the covering is defined as the quotient graph (Y, F) with $Y = H\backslash\mathfrak{T}_{x_0}$ and F the set of edges defined by the canonical projection from $\mathfrak{T}_{x_0}$ onto Y. H can be interpreted as the group of closed geodesics on the quotient graph, based at H_{x_0}, i.e. as the fundamental group of Y.

If H is a normal subgroup, the quotient group (also called the covering group) $H \backslash \Gamma_{x_0}$ acts faithfully on the fiber at x_0. An example is the commutator subgroup $[\Gamma_{x_0}, \Gamma_{x_0}]$. The associate covering is the maximal Abelian covering at x_0.

Exercise 1. Determine the maximal Abelian cover of the tetrahedron.

1.2 Energy

Let us consider a nonnegative function κ on X. Set $\lambda_x = \kappa_x + \sum_y C_{x,y}$ and $P_y^x = \frac{C_{x,y}}{\lambda_x}$. P is a (sub) stochastic transition matrix which is λ-symmetric (i.e. such that $\lambda_x P_y^x = \lambda_y P_x^y$) with $P_x^x = 0$ for all x in X.

It defines a symmetric irreducible Markov chain ξ_n.

We can define above it a continuous time λ-symmetric irreducible Markov chain x_t, with exponential holding times of parameter 1. We have $x_t = \xi_{N_t}$, where N_t denotes a Poisson process of intensity 1. The *infinitesimal generator* is given by $L_y^x = P_y^x - \delta_y^x$.

We denote by P_t its (sub) Markovian semigroup $\exp(Lt) = \sum \frac{t^k}{k!} L^k$. L and P_t are λ-symmetric.

We will use the Markov chain associated with C, κ, sometimes in discrete time, sometimes in continuous time (with exponential holding times).

Recall that for any complex function $z^x, x \in X$, the "energy"

$$e(z) = \langle -Lz, \overline{z} \rangle_\lambda = \sum_{x \in X} -(Lz)^x \overline{z}^x \lambda_x$$

is nonnegative as it can be easily written

$$e(z) = \sum_x \lambda_x z^x \overline{z}^x - \sum_{x,y} C_{x,y} z^x \overline{z}^y = \frac{1}{2} \sum_{x,y} C_{x,y}(z^x - z^y)(\overline{z}^x - \overline{z}^y) + \sum_x \kappa_x z^x \overline{z}^x$$

The Dirichlet space [10] is the space of real functions equipped with the *energy* scalar product

$$e(f,g) = \frac{1}{2} \sum_{x,y} C_{x,y}(f^x - f^y)(g^x - g^y) + \sum_x \kappa_x f^x g^x = \sum_x \lambda_x f^x g^x - \sum_{x,y} C_{x,y} f^x g^y$$

defined by polarization of e.

Note that the non negative symmetric "*conductance matrix*" C and the non negative *equilibrium or "killing" measure* κ are the free parameters of the model.

Exercise 2. Prove that the eigenfunction associated with the lowest eigenvalue of $-L$ is unique and has constant sign by an argument based on the fact that the map $z \to |z|$ lowers the energy (which follows easily from the expression given above).

In quantum mechanics, the infinitesimal generator $-L$ is called the Hamiltonian and its eigenvalues are the energy levels.

One can learn more on graphs and eigenvalues in [2].

We have a dichotomy between:

- The recurrent case where 0 is the lowest eigenvalue of $-L$, and the corresponding eigenspace is formed by constants. Equivalently, $P1 = 1$ and κ vanishes.
- The transient case where the lowest eigenvalue is positive which means there is a "Poincaré inequality": For some positive ε, the energy $e(f, f)$ dominates $\varepsilon \langle f, f \rangle_\lambda$ for all f. Equivalently, as we are on a finite space, κ does not vanish. Note however these equivalences doe not hold in general on infinite spaces, though the dichotomy is still valid.

In the transient case, we denote by V the associated *potential operator* $(-L)^{-1} = \int_0^\infty P_t dt$. It can be expressed in terms of the spectral resolution of L. We will denote $\sum_y V_y^x f^y$ by $(Vf)^x$ or $Vf(x)$.

Note that the function Vf (called the potential of f) is characterized by the identity

$$e(Vf, g) = \langle f, g \rangle_\lambda$$

valid for all functions f and g. The potential operator diverges on positive functions in the recurrent case. These properties define the dichotomy transient/recurrent on infinite spaces.

We denote by G the *Green function* defined on X^2 as $G^{x,y} = \frac{V_y^x}{\lambda_y} = \frac{1}{\lambda_y}[(I - P)^{-1}]_y^x$ i.e. $G = (M_\lambda - C)^{-1}$. It induces a linear bijection from measures into functions. We will denote $\sum_y G^{x,y} \mu_y$ by $(G\mu)^x$ or $G\mu(x)$.

Note that the function $G\mu$ (called the potential of μ) is characterized by the identity

$$e(f, G\mu) = \langle f, \mu \rangle$$

valid for all functions f and measures μ. In particular $G\kappa = 1$ as $e(1, f) = \sum f^x \kappa_x = \langle f, 1 \rangle_\kappa$.

Example 5. The Green function in the case of the complete graph K_n with uniform killing measure of intensity $c > 0$ is given by the matrix

$$\frac{1}{n + c}(I + \frac{1}{c}J)$$

where J denotes the (n, n) matrix with all entries equal to 1.

Proof. Note first that $M_\lambda - C = (n+c)I - J$. The inverse is easily checked. $\square$

See [10] for a development of this theory in a more general setting.

In the recurrent case, the potential operator V can be defined on the space $\lambda^\perp$ of functions f such that $\langle f, 1 \rangle_\lambda = 0$ as the inverse of the restriction of $I - P$ to $\lambda^\perp$. The Green operator G maps the space of measures of total charge zero onto $\lambda^\perp$: setting for any signed measure ν of total charge zero $G\nu = V\frac{\nu}{\lambda}$, we have for any function f, $\langle \nu, f \rangle = e(G\nu, f)$ (as $e(G\nu, 1) = 0$) and in particular $f^x - f^y = e(G(\delta_x - \delta_y), f)$.

Exercise 3. In the case of the complete graph K_n, show that the Green operator is given by:

$$G\nu(x) = \frac{\nu_x}{n}$$

Remark 2. Markov chains with different holding times parameters are associated with the same energy form. If q is any positive function on X, the Markov chain with y_t holding times parameter q^x, $x \in X$ is obtained from x_t by time change: $y_t = x_{\sigma_t}$, where σ_t is the right continuous increasing family of stopping times defined by $\int_0^{\sigma_t} q^{-1}(x_s)ds = t$. Its semigroup is $q^{-1}\lambda$-symmetric with infinitesimal generator given by qL. The potential operator is different but the Green function does not change. In particular, if we set $q^x = \lambda_x$ for all x, the duality measure is the counting measure and the potential operator V is given by the Green function G. The associated rescaled Markov chain will be used in the next chapters.

1.3 Feynman–Kac Formula

A discrete analogue of the Feynman–Kac formula can be given as follows: Let s be any function on X taking values in $(0, 1]$. Then, for the discrete Markov chain ξ_n associated with P, it is a straightforward consequence of the Markov property that:

$$\mathbb{E}_x\left(\prod_{j=0}^{n-1} s(\xi_j) 1_{\{\xi_n = y\}}\right) = [(M_s P)^n]_y^x$$

Similarly, for the continuous time Markov chain x_t (with exponential holding times), we have the Feynman–Kac formula:

Proposition 1. *If $k(x)$ is a nonnegative function defined on X,*

$$\mathbb{E}_x\left(e^{-\int_0^t k(x_s)ds} 1_{\{x_t = y\}}\right) = [\exp(t(L - M_k)]_y^x.$$

Proof. It is enough to check, by differentiating the first member $V(t)$ with respect to t, that $V'(t) = (L - M_k)V(t)$.

Precisely, if we set $(V_t)^x_y = \mathbb{E}_x(e^{-\int_0^t k(x_s)ds}1_{\{x_t=y\}})$, by the Markov property,

$$(V_{t+\Delta t})^x_y = \mathbb{E}_x(e^{-\int_0^t k(x_s)ds}\mathbb{E}_{X_t}(e^{-\int_0^{\Delta t} k(x_s)ds}1_{\{x_{\Delta t}=y\}})$$

$$= \sum_{z\in X}\mathbb{E}_x(e^{-\int_0^t k(x_s)ds}1_{\{x_t=z\}}\mathbb{E}_z(e^{-\int_0^{\Delta t} k(x_s)ds}1_{\{x_{\Delta t}=y\}})$$

$$= \sum_{z\in X}(V_t)^x_z\mathbb{E}_z(e^{-\int_0^{\Delta t} k(x_s)ds}1_{\{x_{\Delta t}=y\}}).$$

Then one verifies easily by considering the first and second times of jump that as Δt goes to zero, $\mathbb{E}_y(e^{-\int_0^{\Delta t} k(x_s)ds}1_{\{x_{\Delta t}=y\}}) - 1$ is equivalent to $-(k(y) + 1)\Delta t$ and for $z \neq y$, $\mathbb{E}_z(e^{-\int_0^{\Delta t} k(x_s)ds}1_{\{x_{\Delta t}=y\}})$ is equivalent to $P^z_y\Delta t$.   □

Exercise 4. Verify the last assertion of the proof.

For any nonnegative measure χ, set $V_\chi = (-L + M_{\frac{\chi}{\lambda}})^{-1}$ and $G_\chi = V_\chi M_{\frac{1}{\lambda}} = (M_\lambda + M_\chi - C)^{-1}$. It is a symmetric nonnegative function on $X \times X$. G_0 is the Green function G, and G_χ can be viewed as the Green function of the energy form $e_\chi = e + \|\ \ \|^2_{L^2(\chi)}$.

Note that e_χ has the same conductances C as e, but χ is added to the killing measure. Note also that V_χ is not the potential of the Markov chain associated with e_χ when one takes exponential holding times of parameter 1: the holding time expectation at x becomes $\frac{1}{1+\chi(x)}$. But the Green function is intrinsic i.e. invariant under a change of time scale. Still, we have by Feynman Kac formula

$$\int_0^\infty \mathbb{E}_x(e^{-\int_0^t \frac{\chi}{\lambda}(x_s)ds}1_{\{x_t=y\}})dt = [V_\chi]^x_y.$$

We have also the "generalized resolvent equation" $V - V_\chi = V M_{\frac{\chi}{\lambda}}V_\chi = V_\chi M_{\frac{\chi}{\lambda}}V$. Then,

$$G - G_\chi = GM_\chi G_\chi = G_\chi M_\chi G \tag{1.1}$$

Exercise 5. Prove the generalized resolvent equation.

Note that the recurrent Green operator G defined on signed measures of zero charge is the limit of the transient Green operator G_χ, as $\chi \to 0$.

1.4   Recurrent Extension of a Transient Chain

It will be convenient to add a cemetery point Δ to X, and extend C, λ and G to $X^\Delta = \{X \cup \Delta\}$ by setting, $\lambda_\Delta = \sum_{x\in X}\kappa_x$, $C_{x,\Delta} = \kappa_x$ and $G^{x,\Delta} = G^{\Delta,x} = G^{\Delta,\Delta} = 0$ for all $x \in X$. Note that $\lambda(X^\Delta) = \sum_{X\times X}C_{x,y} + 2\sum_X \kappa_x = \lambda(X) + \lambda_\Delta$.

One can consider the recurrent "resurrected" Markov chain defined by the extensions of the conductances to X^Δ. An energy e^Δ is defined by the formula

$$e^\Delta(z) = \frac{1}{2} \sum_{x,y \in X^\Delta} C_{x,y}(z^x - z^y)(\overline{z}^x - \overline{z}^y)$$

From the irreducibility assumption, it follows that e^Δ vanishes only on constants. We denote by P^Δ the transition kernel on X^Δ defined by

$$[P^\Delta]_y^x = \frac{C_{x,y}}{\sum_{y \in X^\Delta} C_{x,y}} = \frac{C_{x,y}}{\lambda_x}$$

Note that $P^\Delta 1 = 1$ so that λ is now an invariant measure with $\lambda_x [P^\Delta]_y^x = \lambda_y [P^\Delta]_x^y$ on X^Δ. Also

$$e^\Delta(f,g) = \langle f - P^\Delta f, g \rangle_\lambda$$

Denote V^Δ and G^Δ the associated potential and Green operators.

Note that for μ carried by X, for all $x \in X$, denoting by ε_Δ the unit point mass at Δ,

$$\mu_x = e^\Delta(G^\Delta(\mu - \mu(X)\varepsilon_\Delta), 1_x) = \lambda_x((I - P^\Delta)G^\Delta(\mu - \mu(X)\varepsilon_\Delta)(x)$$

$$= \lambda_x((I - P)G^\Delta(\mu - \mu(X)\varepsilon_\Delta))(x) - \kappa_x G^\Delta(\mu - \mu(X)\varepsilon_\Delta)(\Delta).$$

Hence, applying G, it follows that on X^Δ,

$$G\mu - G^\Delta(\mu - \mu(X)\varepsilon_\Delta) - G^\Delta(\mu - \mu(X)\varepsilon_\Delta)(\Delta)G\kappa = G^\Delta(\mu - \mu(X)\varepsilon_\Delta)$$
$$- G^\Delta(\mu - \mu(X)\varepsilon_\Delta)(\Delta).$$

Moreover, as $G^\Delta(\mu - \mu(X)\varepsilon_\Delta)$ is in $\lambda^\perp$, integrating by λ, we obtain that

$$\sum_{x \in X} \lambda_x G(\mu)^x = -G^\Delta(\mu - \mu(X)\varepsilon_\Delta)(\Delta)\lambda(X^\Delta).$$

Therefore, $G^\Delta(\mu - \mu(X)\varepsilon_\Delta)(\Delta) = \frac{-\langle \lambda, G\mu \rangle}{\lambda(X^\Delta)}$ and we get the following:

Proposition 2. *For any measure μ on X,* $G^\Delta(\mu - \mu(X)\varepsilon_\Delta) = -\frac{\langle \lambda, G\mu \rangle}{\lambda(X^\Delta)} + G\mu.$

This type of extension can be done in a more general context (See [25] and Dellacherie–Meyer [4])

Remark 3. Conversely, a recurrent chain can be killed at any point x_0 of X, defining a Green function $G^{X-\{x_0\}}$ on $X - \{x_0\}$. Then, for any μ carried by $X - \{x_0\}$,

$$G^{X-\{x_0\}}\mu = G(\mu - \mu(X)\varepsilon_{x_0}) - G(\mu - \mu(X)\varepsilon_{x_0})(x_0).$$

This transient chain allows to recover the recurrent one by the above procedure.

Exercise 6. Consider a transient process which is killed with probability p at each passage in Δ. Determine the associated energy and Green operator.

1.5   Transfer Matrix

Let us suppose in this section that we are in the *recurrent case*: We can define a scalar product on the space $\mathbb{A}$ of functions on E^o (oriented edges) as follows $\langle \omega, \eta \rangle_{\mathbb{A}} = \frac{1}{2} \sum_{x,y} C_{x,y} \omega^{x,y} \eta^{x,y}$. Denoting as in [29] $df^{u,v} = f^v - f^u$, we note that $\langle df, dg \rangle_{\mathbb{A}} = e(f,g)$. In particular

$$\langle df, dG(\delta_y - \delta_x) \rangle_{\mathbb{A}} = df^{x,y}$$

Denote $\mathbb{A}_-$, ($\mathbb{A}_+$) the space of real valued functions on E^o odd (even) for orientation reversal. Note that the spaces $\mathbb{A}_+$ and $\mathbb{A}_-$ are orthogonal for the scalar product defined on $\mathbb{A}$. The space $\mathbb{A}_-$ should be viewed as the space of "discrete differential forms".

Following this analogy, define for any α in $\mathbb{A}_-$, define $d^*\alpha$ by $(d^*\alpha)^x = -\sum_{y \in X} P_y^x \alpha^{x,y}$. Note it belongs to $\lambda^\perp$ as $\sum_{x,y} C_{x,y}\alpha^{x,y}$ vanishes. We have

$$\langle \alpha, df \rangle_{\mathbb{A}} = \frac{1}{2}\sum_{x,y} \lambda_x P_y^x \alpha^{x,y}(f^y - f^x)$$

$$= \frac{1}{2}\sum_{x \in X}(d^*\alpha)^x f^x \lambda_x - \frac{1}{2}\sum_{x,y} \lambda_x P_y^x \alpha^{y,x} f^y = \sum_{x \in X}(d^*\alpha)^x f^x \lambda_x$$

as the two terms of the difference are in fact opposite since α is skew symmetric. The image of d and the kernel of d^* are therefore orthogonal in $\mathbb{A}_-$. We say α in $\mathbb{A}_-$ is harmonic iff $d^*\alpha = 0$.

Moreover,

$$e(f,f) = \langle df, df \rangle_{\mathbb{A}} = \sum_{x \in X}(d^*df)^x f^x \lambda_x.$$

Note also that for any function f,

$$d^*df = -Pf + f = -Lf.$$

d is the discrete analogue of the differential and d^* the analogue of its adjoint, depending on the metric which is here defined by the conductances. L is a discrete version of the Laplacian.

Proposition 3. *The projection of any α in $\mathbb{A}_-$ on the image of d is $dVd^*(\alpha)$.*

Proof. Indeed, for any function g, $\langle \alpha, dg \rangle_{\mathbb{A}} = \langle d^*\alpha, g \rangle_\lambda = e(Vd^*\alpha, g) = \langle dVd^*(\alpha), dg \rangle_{\mathbb{A}}$. $\square$

We now can come to the definition of the transfer matrix: Set $\alpha^{x,y}_{(u,v)} = \pm\frac{1}{C_{u,v}}$ if $(x,y) = \pm(u,v)$ and 0 elsewhere. Then $\lambda_x d^*\alpha_{(u,v)}(x) = \delta^x_v - \delta^x_u$ and $dVd^*(\alpha_{(u,v)}) = dG(\delta_v - \delta_u)$. Note that given any orientation of the graph, the family $\{\alpha^*_{(u,v)} = \sqrt{C_{u,v}}\alpha_{(u,v)}, (u,v) \in E^+\}$ is an orthonormal basis of $\mathbb{A}_-$ (here E^+ denotes the set of positively oriented edges).

The symmetric transfer matrix $K^{(x,y),(u,v)}$, indexed by pairs of oriented edges, is defined to be

$$K^{(x,y),(u,v)} = [dG(\delta_v - \delta_u)]^{x,y} = G(\delta_v - \delta_u)^y - G(\delta_v - \delta_u)^x$$

$$=< dG(\delta_y - \delta_x), dG(\delta_v - \delta_u) >_{\mathbb{A}}$$

for $x, y, u, v \in X$, with $C_{x,y}C_{u,v} > 0$.

As $dG((d^*\alpha)\lambda) = dVd^*(\alpha)$ is the projection $\Pi(\alpha)$ of α on the image of d in $\mathbb{A}_-$, we have also:

$$< \alpha_{(u,v)}, \Pi(\alpha_{(u',v')}) >_{\mathbb{A}}=< \alpha_{(u,v)}, dG(\delta_{v'} - \delta_{u'}) >_{\mathbb{A}}= K^{(u,v),(u',v')}$$

For every oriented edge $h = (x, y)$ in X, set $K^h = dG(\delta^y - \delta^x)$. We have $\langle K^h, K^g \rangle_{\mathbb{A}} = K^{h,g}$. We can view dG as a linear operator mapping the space measures of total charge zero into $\mathbb{A}_-$. As measures of the form $\delta_y - \delta_x$ span the space of measures of total charge zero, it is determined by the transfer matrix.

Note that $d^*dGv = v/\lambda$ for any v of total charge zero and that for all α in $\mathbb{A}_-$, $(d^*\alpha)\lambda$ has total charge zero.

Consider now, in the transient case, the transfer matrix associated with G^Δ.

We see that for x and y in X, $G^\Delta(\delta_x - \delta_y)^u - G^\Delta(\delta_x - \delta_y)^v = G(\delta_x - \delta_y)^u - G(\delta_x - \delta_y)^v$.

We can see also that $G^\Delta(\delta_x - \delta_\Delta) = G\delta_x - \frac{\langle \lambda, G\delta_x \rangle}{\lambda(X^\Delta)}$. So the same identity holds in X^Δ.

Therefore, as $G^{x,\Delta} = 0$, in all cases,

$$K^{(x,y),(u,v)} = G^{x,u} + G^{y,v} - G^{x,v} - G^{y,u}$$

Exercise 7. Cohomology and complex transition matrices.

Consider, in the recurrent case, $\omega \in \mathbb{A}_-$ such that $d^*\omega = 0$. Note that the space H^1 of such ω's is isomorphic to the first cohomology space, defined as the quotient $\mathbb{A}_- / \operatorname{Im}(d)$. Prove that $P(I + i\omega)$ is λ-self adjoint on X, maps 1 onto 1 and that we have $\mathbb{E}_x(\prod_{j=0}^{n-1}(1+\omega(\xi_j, \xi_{j+1}))1_{\{\xi_n = y\}}) = [(P(I+i\omega))^n]_y^x$.

Chapter 2
Loop Measures

2.1 A Measure on Based Loops

We denote by $\mathbb{P}^x$ the family of probability laws on piecewise constant paths defined by P_t.

$$\mathbb{P}^x(\gamma(t_1) = x_1, \ldots, \gamma(t_h) = x_h) = P_{t_1}(x, x_1) P_{t_2 - t_1}(x_1, x_2) \ldots P_{t_h - t_{h-1}}(x_{h-1}, x_h)$$

The corresponding process is a Markov chain in continuous time. It can also be constructed as the process ξ_{N_t}, where ξ_n is the discrete time Markov chain starting at x, with transition matrix P, and N_t an independent Poisson process.

In the transient case, the lifetime is a.s. finite and denoting by $p(\gamma)$ the number of jumps and T_i the jump times, we have:

$$\mathbb{P}^x(p(\gamma) = k, \gamma_{T_1} = x_1, \ldots, \gamma_{T_k} = x_k, T_1 \in dt_1, \ldots, T_k \in dt_k)$$

$$= \frac{C_{x,x_1} \ldots C_{x_{k-1},x_k} \kappa_{x_k}}{\lambda_x \lambda_{x_1} \ldots \lambda_{x_k}} 1_{\{0 < t_1 < \ldots < t_k\}} e^{-t_k} dt_1 \ldots dt_k$$

For any integer $p \geq 2$, let us define a *based loop* with p points in X as a couple $l = (\xi, \tau) = ((\xi_m, 1 \leq m \leq p), (\tau_m, 1 \leq m \leq p+1))$ in $X^p \times \mathbb{R}_+^{p+1}$, and set $\xi_{p+1} = \xi_1$ (equivalently, we can parametrize the associated discrete based loop by $\mathbb{Z}/p\mathbb{Z}$). The integer p represents the number of points in the discrete based loop $\xi = (\xi_1, \ldots, \xi_{p(\xi)})$ and will be denoted $p(\xi)$, and the τ_m are holding times. Note however that two time parameters are attached to the base point since the based loops do not in general end or start with a jump.

Based loops with one point ($p = 1$) are simply given by a pair (ξ, τ) in $X \times \mathbb{R}_+$.

Y. Le Jan, *Markov Paths, Loops and Fields*, Lecture Notes in Mathematics 2026, DOI 10.1007/978-3-642-21216-1_2, © Springer-Verlag Berlin Heidelberg 2011

Based loops have a natural time parametrization $l(t)$ and a time period $T(\xi) = \sum_{i=1}^{p(\xi)+1} \tau_i$. If we denote $\sum_{i=1}^{m} \tau_i$ by T_m: $l(t) = \xi_m$ on $[T_{m-1}, T_m]$ (with by convention $T_0 = 0$ and $\xi_1 = \xi_{p+1}$).

Let $\mathbb{P}_t^{x,y}$ denote the (non normalized) "bridge measure" on piecewise constant paths from x to y of duration t constructed as follows:

If $t_1 < t_2 < \dots < t_h < t$,

$$\mathbb{P}_t^{x,y}(l(t_1) = x_1, \dots, l(t_h) = x_h) = [P_{t_1}]_{x_1}^{x_h}[P_{t_2-t_1}]_{x_2}^{x_1}\dots[P_{t-t_h}]_y^{x_h}\frac{1}{\lambda_y}$$

Its mass is $p_t^{x,y} = \frac{[P_t]_y^x}{\lambda_y}$. For any measurable set A of piecewise constant paths indexed by $[0\ t]$, we can also write

$$\mathbb{P}_t^{x,y}(A) = \mathbb{P}_x(A \cap \{x_t = y\})\frac{1}{\lambda_y}.$$

Exercise 8. Prove that $\mathbb{P}_t^{y,x}$ is the image of $\mathbb{P}_t^{x,y}$ by the operation of time reversal on paths indexed by $[0\ t]$.

A σ-finite measure μ is defined on based loops by

$$\mu = \sum_{x \in X} \int_0^\infty \frac{1}{t} \mathbb{P}_t^{x,x} \lambda_x dt$$

Remark 4. The introduction of the factor $\frac{1}{t}$ will be justified in the following. See in particular formula (2.3). It can be interpreted as the normalization of the uniform measure on the loop, according to which the base point is chosen.

From the expression of the bridge measure, we see that by definition of μ, if $t_1 < t_2 < \dots < t_h < t$,

$$\mu(l(t_1) = x_1, \dots, l(t_h) = x_h, T \in dt) = [P_{t_1+t-t_h}]_{x_1}^{x_h}[P_{t_2-t_1}]_{x_2}^{x_1}\dots[P_{t_h-t_{h-1}}]_{x_h}^{x_{h-1}}\frac{1}{t}dt.$$
$$(2.1)$$

Note also that for $k > 1$, using the second expression of $\mathbb{P}_t^{x,y}$ and the fact that conditionally on $N_t = k$, the jump times are distributed like an increasingly reordered k-uniform sample of $[0\ t]$

$$\lambda_x \mathbb{P}_t^{x,x}(p = k, \xi_1 = x_1, \xi_2 = x_2, \dots, \xi_k = x_k, T_1 \in dt_1, \dots, T_k \in dt_k)$$

$$= 1_{\{x=x_1\}}e^{-t}\frac{t^k}{k!}P_{x_2}^{x_1}P_{x_3}^{x_2}\dots P_{x_1}^{x_k}1_{\{0<t_1<\dots t_k<t\}}\frac{k!}{t^k}dt_1\dots dt_k$$

$$= 1_{\{x=x_1\}}P_{x_2}^{x_1}P_{x_3}^{x_2}\dots P_x^{x_k}1_{\{0<t_1<\dots t_k<t\}}e^{-t}dt_1\dots dt_k$$

Therefore,

$$\mu(p = k, \xi_1 = x_1, .., \xi_k = x_k, T_1 \in dt_1, .., T_k \in dt_k, T \in dt) \qquad (2.2)$$

$$= P_{x_2}^{x_1} .. P_{x_1}^{x_k} \frac{1_{\{0 < t_1 < ... < t_k < t\}}}{t} e^{-t} dt_1 ... dt_k dt \qquad (2.3)$$

for $k > 1$.

Moreover, for one point-loops, $\mu\{p(\xi) = 1, \xi_1 = x_1, \tau_1 \in dt\} = \frac{e^{-t}}{t} dt$.

It is clear on these formulas that for any positive constant c, the energy forms e and ce define the same loop measure.

2.2 First Properties

Note that the loop measure is invariant under time reversal.

If D is a subset of X, the restriction of μ to loops contained in D, denoted μ^D is clearly the loop measure induced by the Markov chain killed at the exit of D. This can be called the *restriction property*.

Let us recall that this killed Markov chain is defined by the restriction of λ to D and the restriction P^D of P to D^2 (or equivalently by the restriction e_D of the Dirichlet form e to functions vanishing outside D).

As $\int \frac{t^{k-1}}{k!} e^{-t} dt = \frac{1}{k}$, it follows from (2.2) that for $k > 1$, on based loops,

$$\mu(p(\xi) = k, \xi_1 = x_1, \ldots, \xi_k = x_k) = \frac{1}{k} P_{x_2}^{x_1} ... P_{x_1}^{x_k}. \qquad (2.4)$$

In particular, we obtain that, for $k \geq 2$

$$\mu(p = k) = \frac{1}{k} Tr(P^k)$$

and therefore, as $Tr(P) = 0$, in the transient case:

$$\mu(p > 1) = \sum_2^\infty \frac{1}{k} Tr(P^k) = -\log(\det(I - P)) = \log(\det(G) \prod_x \lambda_x) \qquad (2.5)$$

since (denoting M_λ the diagonal matrix with entries λ_x), we have

$$\det(I - P) = \frac{\det(M_\lambda - C)}{\det(M_\lambda)}$$

Note that $\det(G)$ is defined as the determinant of the matrix $G^{x,y}$. It is the determinant of the matrix representing the scalar product defined on $\mathbb{R}^{|X|}$ (more precisely, on the space of measures on X) by G in any basis, orthonormal with respect to the natural euclidean scalar product on $\mathbb{R}^{|X|}$.

Moreover

$$\int p(l)1_{\{p>1\}}\mu(dl) = \sum_{2}^{\infty} Tr(P^k) = Tr((I-P)^{-1}P) = Tr(GC)$$

2.3   Loops and Pointed Loops

It is clear on formula (2.1) that μ is invariant under the time shift that acts
naturally on based loops.

A loop is defined as an equivalence class of based loops for this shift.
Therefore, μ induces a *measure on loops also denoted by* μ.

A loop is defined by the discrete loop ξ° formed by the ξ_i in circular order,
(i.e. up to translation) and the associated holding times. We clearly have:

$$\mu(\xi^{\circ} = (x_1, x_2, \ldots, x_k)^{\circ}) = P_{x_2}^{x_1}...P_{x_1}^{x_k}$$

provided the loop is primitive i.e. does not have a non trivial period, as it is
in this case formed by p equivalent based loops. Otherwise, the right hand
side should be divided by the mutiplicity. However, loops are not easy to
parametrize, that is why we will work mostly with based loops or with *pointed
loops*. These are defined as based loops ending with a jump, or equivalently
as loops with a starting point. They can be parametrized by a based discrete
loop and by the holding times at each point. Calculations are easier if we work
with based or pointed loops, even though we will deal only with functions
independent of the base point.

The parameters of the pointed loop naturally associated with a based loop
are $\xi_1, \ldots, \xi_p$ and

$$\tau_1 + \tau_{p+1} = \tau_1^*, \tau_i = \tau_i^*,\ 2 \le i \le p$$

An elementary change of variables, shows the expression of μ on pointed loops
can be written:

$$\mu(p = k, \xi_i = x_i, \tau_i^* \in dt_i) = P_{x_2}^{x_1}...P_{x_1}^{x_k} \frac{t_1}{\sum t_i} e^{-\sum t_i} dt_1...dt_k. \qquad (2.6)$$

Trivial ($p = 1$) pointed loops and trivial based loops coincide.

Note that loop functionals can be written

$$\Phi(l^{\circ}) = \sum 1_{\{p=k\}} \Phi_k((\xi_i, \tau_i^*), i = 1, \ldots, k)$$

with Φ_k invariant under circular permutation of the variables (ξ_i, τ_i^*).

Then, for non negative Φ_k

$$\int \Phi_k(l^{\circ})\mu(dl) = \sum \int \Phi_k((x_i,t_i)i = 1,\ldots,k)P_{x_2}^{x_1}\ldots P_{x_1}^{x_k}e^{-\sum t_i}\frac{t_1}{\sum t_i}dt_1\ldots dt_k$$

and by invariance under circular permutation, the term t_1 can be replaced by any t_i. Therefore, adding up and dividing by k, we get that

$$\int \Phi_k(l^{\circ})\mu(dl) = \sum \int \frac{1}{k}\Phi_k((x_i,t_i)i = 1,\ldots,k)P_{x_2}^{x_1}\ldots P_{x_1}^{x_k}e^{-\sum t_i}dt_1\ldots dt_k.$$

The expression on the right side, applied to any pointed loop functional defines a different measure on pointed loops, we will denote by μ^*. It induces the same measure as μ on loops.

We see on this expression that conditionally on the discrete loop, the holding times of the loop are independent exponential variables.

$$\mu^*(p = k, \xi_i = x_i, \tau_i^* \in dt_i) = \frac{1}{k}\prod_{i\in\mathbb{Z}/p\mathbb{Z}}\frac{C_{x_i,x_{i+1}}}{\lambda_{x_i}}e^{-t_i}dt_i \qquad (2.7)$$

Conditionally on $p(\xi) = k$, T is a gamma variable of density $\frac{t^{k-1}}{(k-1)!}e^{-t}$ on $\mathbb{R}_+$ and $(\frac{\tau_i^*}{T}, 1 \leq i \leq k)$ an independent ordered k-sample of the uniform distribution on $(0,T)$ (whence the factor $\frac{1}{t}$). Both are independent, conditionally on the number of points p of the discrete loop. We see that μ on based loops is obtained from μ on the loops by choosing the base point uniformly. On the other hand, it induces a choice of ξ_1 biased by the size of the τ_i^*'s, different from μ^* for which this choice is uniform (whence the factor $\frac{1}{k}$). But we will consider only loop functionals for which μ and μ^* coincide.

It will be convenient to rescale the holding time at each ξ_i by λ_{ξ_i} and set

$$\widehat{\tau}_i = \frac{\tau_i^*}{\lambda_{\xi_i}}.$$

The discrete part of the loop is the most important, though we will see that to establish a connection with Gaussian fields it is necessary to consider occupation times. The simplest variables are the number of jumps from x to y, defined for every oriented edge (x,y)

$$N_{x,y} = \#\{i : \xi_i = x, \xi_{i+1} = y\}$$

(recall the convention $\xi_{p+1} = \xi_1$) and

$$N_x = \sum_y N_{x,y}$$

Note that $N_x = \#\{i \geq 1 : \xi_i = x\}$ except for trivial one point loops for which it vanishes.

Then, the measure on pointed loops (2.6) can be rewritten as:

$$\mu^*(p = 1, \xi = x, \hat{\tau} \in dt) = e^{-\lambda_x t}\frac{dt}{t} \text{ and} \tag{2.8}$$

$$\mu^*(p = k, \xi_i = x_i, \hat{\tau}_i \in dt_i) = \frac{1}{k}\prod_{x,y} C_{x,y}^{N_{x,y}}\prod_x \lambda_x^{-N_x}\prod_{i\in\mathbb{Z}/p\mathbb{Z}}\lambda_{\xi_i}e^{-\lambda_{\xi_i}t_i}dt_i. \tag{2.9}$$

Another *bridge measure* $\mu^{x,y}$ can be defined on paths γ from x to y:

$$\mu^{x,y}(d\gamma) = \int_0^\infty \mathbb{P}_t^{x,y}(d\gamma)dt.$$

Note that the mass of $\mu^{x,y}$ is $G^{x,y}$. We also have, with similar notations as the one defined for loops, p denoting the number of jumps

$$\mu^{x,y}(p(\gamma) = k, \gamma_{T_1} = x_1, \ldots, \gamma_{T_{k-1}} = x_{k-1}, T_1 \in dt_1, \ldots, T_k \in dt_k, T \in dt)$$

$$= \frac{C_{x,x_1}C_{x_1,x_2}\ldots C_{x_{k-1},y}}{\lambda_x\lambda_{x_1}\ldots\lambda_y}1_{\{0<t_1<\ldots<t_k<t\}}e^{-t}dt_1\ldots dt_k dt.$$

From now on, we will assume, unless otherwise specified, that we are in the *transient case*.

For any $x \neq y$ in X and $s \in [0,1]$, setting $P_v^{(s),u} = P_v^u$ if $(u,v) \neq (x,y)$ and $P_y^{(s),x} = sP_y^x$, we can prove in the same way as (2.5) that:

$$\mu(s^{N_{x,y}}1_{\{p>1\}}) = -\log(\det(I - P^{(s)})).$$

Differentiating in $s = 1$, and remembering that for any invertible matrix function $M(s)$, $\frac{d}{ds}\log(\det(M(s)) = Tr(M'(s)M(s)^{-1})$, it follows that:

$$\mu(N_{x,y}) = [(I - P)^{-1}]_x^y P_y^x = G^{x,y}C_{x,y}$$

and

$$\mu(N_x) = \sum_y \mu(N_{x,y}) = \lambda_x G^{x,x} - 1 \tag{2.10}$$

(as $G(M_\lambda - C) = Id$).

Exercise 9. Show that more generally

$$\mu(N_{x,y}(N_{x,y} - 1)\ldots(N_{x,y} - k + 1)) = (k-1)!(G^{x,y}C_{x,y})^k.$$

Hint: Show that if $M''(s)$ vanishes,

$$\frac{d^n}{ds^n}\log(\det(M(s))) = (-1)^{n-1}(n-1)!Tr((M'(s)M(s)^{-1})^n).$$

Exercise 10. Show that more generally, if x_i, y_i are n distinct oriented edges:

$$\mu(\prod N_{x_i,y_i}) = \prod C_{x_i,y_i} \frac{1}{n} \sum_{\sigma \in S_n} \prod G^{y_{\sigma(i)},x_{\sigma(i+1)}}$$

Hint: Introduce $[P^{(s_1,\ldots,s_n)}]^x_y$ equal to P^x_y if $(x,y) \neq (x_i, y_i)$ for all i, and equal to $s_i P^{x_i}_{y_i}$ if $(x,y) = (x_i, y_i)$.

We finally note that if $C_{x,y} > 0$, any path segment on the graph starting at x and ending at y can be naturally extended into a loop by adding a jump from y to x. We have the following

Proposition 4. For $C_{x,y} > 0$, the natural extension of $\mu^{x,y}$ to loops coincides with $\frac{N_{y,x}(l)}{C_{x,y}} \mu(dl)$.

Proof. The first assertion follows from the formulas, noticing that a loop l can be associated to $N_{y,x}(l)$ distinct bridges from x to y, obtained by "cutting" one jump from y to x. $\qquad\square$

Note that a) shows that the loop measure induces bridge measures $\mu^{x,y}$ when $C_{x,y} > 0$. If $C_{x,y}$ vanishes, an arbitrarily small positive perturbation creating a non vanishing conductance between x and y allows to do it. More precisely, denoting by $e^{(\varepsilon)}$ the energy form equal to e except for the additional conductance $C^{(\varepsilon)}_{x,y} = \varepsilon$, $\mu^{x,y}$ can be represented as $\frac{d}{d\varepsilon} \mu^{e^{(\varepsilon)}}|_{\varepsilon=0}$.

2.4 Occupation Field

To each loop l° we associate local times, i.e. an occupation field $\{\widehat{l}_x, x \in X\}$ defined by

$$\widehat{l}^x = \int_0^{T(l)} 1_{\{l(s)=x\}} \frac{1}{\lambda_{l(s)}} ds = \sum_{i=1}^{p(l)} 1_{\{\xi_i = x\}} \widehat{\tau}_i$$

for any representative $l = (\xi_i, \tau_i^*)$ of l°.

For a path γ, $\widehat{\gamma}$ is defined in the same way.

Note that

$$\mu((1 - e^{-\alpha \widehat{l}^x}) 1_{\{p=1\}}) = \int_0^\infty e^{-t}(1 - e^{-\frac{\alpha}{\lambda_x} t}) \frac{dt}{t} = \log(1 + \frac{\alpha}{\lambda_x}). \qquad (2.11)$$

The proof goes by expanding $1 - e^{-\frac{\alpha}{\lambda_x} t}$ before the integration, assuming first that α is small and then by analyticity of both members, or more elegantly, noticing that $\int_a^b (e^{-cx} - e^{-dx}) \frac{dx}{x}$ is symmetric in (a,b) and (c,d), by Fubini's theorem.

In particular, $\mu(\widehat{l}^x 1_{\{p=1\}}) = \frac{1}{\lambda_x}$.

From formula (2.7), we get easily that the joint conditional distribution of $(\widehat{l^x},\ x \in X)$ given $(N_x,\ x \in X)$ is a product of gamma distributions. In particular, from the expression of the moments of a gamma distribution, we get that for any function Φ of the discrete loop and $k \geq 1$,

$$\mu((\widehat{l^x})^k 1_{\{p>1\}}\Phi) = \lambda_x^{-k}\mu((N_x + k - 1)...(N_x + 1)N_x\Phi).$$

In particular, by (2.10) $\mu(\widehat{l^x}) = \frac{1}{\lambda_x}[\mu(N_x) + 1] = G^{x,x}$.

Note that functions of $\widehat{l}$ are not the only functions naturally defined on the loops. Other such variables of interest are, for $n \geq 2$, the multiple local times, defined as follows:

$$\widehat{l^{x_1,...,x_n}} = \sum_{j=0}^{n-1} \int_{0<t_1<...<t_n<T} 1_{\{l(t_1)=x_{1+j},...,l(t_{n-j})=x_n,...,l(t_n)=x_j\}} \prod \frac{1}{\lambda_{x_i}}dt_i.$$

It is easy to check that, when the points x_i are distinct,

$$\widehat{l^{x_1,...,x_n}} = \sum_{j=0}^{n-1} \sum_{1\leq i_1<...<i_n\leq p(l)} \prod_{l=1}^{n} 1_{\{\xi_{i_l}=x_{l+j}\}}\widehat{\tau_{i_l}}. \tag{2.12}$$

Note that in general $\widehat{l^{x_1,...,x_k}}$ cannot be expressed in terms of $\widehat{l}$, but

$$\widehat{l^{x_1}}...\widehat{l^{x_n}} = \frac{1}{n} \sum_{\sigma \in \mathcal{S}_n} \widehat{l^{x_{\sigma(1)},...,x_{\sigma(n)}}}.$$

In particular, $\widehat{l^{x,...,x}} = \frac{1}{(n-1)!}[\widehat{l^x}]^n$. It can be viewed as a n-th self intersection local time.

One can deduce from the definitions of μ the following:

Proposition 5. $\mu(\widehat{l^{x_1,...,x_n}}) = G^{x_1,x_2}G^{x_2,x_3}...G^{x_n,x_1}$.

In particular, $\mu(\widehat{l^{x_1}}...\widehat{l^{x_n}}) = \frac{1}{n}\sum_{\sigma \in \mathcal{S}_n}G^{x_{\sigma(1)},x_{\sigma(2)}}G^{x_{\sigma(2)},x_{\sigma(3)}}...G^{x_{\sigma(n)},x_{\sigma(1)}}$.

Proof. Let us denote $\frac{1}{\lambda_y}[P_t]^x_y$ by $p_t^{x,y}$ or $p_t(x,y)$. From the definition of $\widehat{l^{x_1,...,x_n}}$ and μ, $\mu(\widehat{l^{x_1,...,x_n}})$ equals:

$$\sum_x \lambda_x \sum_{j=0}^{n-1} \int\int_{\{0<t_1<...<t_n<t\}} \frac{1}{t}p_{t_1}(x,x_{1+j})\ldots p_{t-t_n}(x_{n+j},x) \prod dt_i dt.$$

where sums of indices $k+j$ are computed $\mathrm{mod}(n)$. By the semigroup property, it equals

$$\sum_{j=0}^{n-1} \int\int_{\{0<t_1<...<t_n<t\}} \frac{1}{t}p_{t_2-t_1}(x_{1+j},x_{2+j})\ldots p_{t_1+t-t_n}(x_{n+j},x_{1+j}) \prod dt_i dt.$$

Performing the change of variables $v_2 = t_2 - t_1, .., v_n = t_n - t_{n-1}, v_1 = t_1 + t - t_n$, and $v = t_1$, we obtain:

$$\sum_{j=0}^{n-1} \int_{\{0<v<v_1, 0<v_i\}} \frac{1}{v_1 + ... + v_n} p_{v_2}(x_{1+j}, x_{2+j}) \ldots p_{v_1}(x_{n+j}, x_{1+j}) \prod dv_i dv$$

$$= \sum_{j=0}^{n-1} \int_{\{0<v_i\}} \frac{v_1}{v_1 + ... + v_n} p_{v_2}(x_{1+j}, x_{2+j}) \ldots p_{v_1}(x_{n+j}, x_{1+j}) \prod dv_i$$

$$= \sum_{j=1}^{n} \int_{\{0<v_i\}} \frac{v_j}{v_1 + ... + v_n} p_{v_2}(x_1, x_2) \ldots p_{v_1}(x_n, x_1) \prod dv_i$$

$$= \int_{\{0<v_i\}} p_{v_2}(x_1, x_2) \ldots p_{v_1}(x_n, x_1) \prod dv_i$$

$$= G^{x_1, x_2} G^{x_2, x_3} ... G^{x_n, x_1}.$$

Note that another proof can be derived from formula (2.12) . $\qquad \square$

Exercise 11. (Shuffle product) Given two positive integers $n > k$, let $\mathcal{P}_{n,k}$ be the family of partitions of $\{1, 2, ...n\}$ into k consecutive non empty intervals $I_l = (i_l, i_l + 1, \ldots, i_{l+1} - 1)$ with $i_1 = 1 < i_2 < ... < i_k < i_{k+1} = n + 1$. Show that

$$\widehat{l}^{x_1, ..., x_n} \widehat{l}^{y_1, ..., y_m} = \sum_{j=0}^{m-1} \sum_{k=1}^{\inf(n,m)} \sum_{I \in \mathcal{P}_{n,k}} \sum_{J \in \mathcal{P}_{m,k}} \widehat{l}^{x_{I_1}, y_{j+J_1}, x_{I_2}, ..., y_{j+J_k}}$$

where for example the term y_{j+J_1} appearing in the upper index should be read as $j + j_1, \ldots, j + j_2 - 1$.

Similarly, we can define $N_{(x_1, y_1), ... (x_n, y_n)}$ to be

$$\sum_{j=0}^{n-1} \sum_{1 \le i_1 < ... < i_n \le p(l)} \prod_{l=1}^{n} 1_{\{\xi_{i_l} = x_{l+j}, \xi_{i_l+1} = y_{l+j}\}}.$$

If $(x_i, y_i) = (x, y)$ for all i, it equals $\frac{N_{x,y}(N_{x,y}-1)...(N_{x,y}-n+1)}{(n-1)!}$.
Notice that

$$\prod N_{(x_i, y_i)} = \frac{1}{n} \sum_{\sigma \in S_n} N_{(x_{\sigma(1)}, y_{\sigma(1)}), ... (x_{\sigma(n)}, y)}.$$

Then we have the following:

Proposition 6. $\int N_{(x_1,y_1),...,(x_n,y_n)}(l)\mu(dl) = \left(\prod C_{x_i,y_i}\right)G^{y_1,x_2}G^{y_2,x_3}\cdots$
G^{y_n,x_1}.

The proof is left as exercise.

Exercise 12. For $x_1 = x_2 = ... = x_k$, we could define different self intersection local times

$$\widehat{l}^{x,(k)} = \sum_{1\le i_1<..<i_k\le p(l)} \prod_{l=1}^{k} 1_{\{\xi_{i_l}=x\}}\widehat{\tau_{i_l}}$$

which vanish on $N_x < k$. Note that

$$\widehat{l}^{x,(2)} = \frac{1}{2}\left((\widehat{l}^x)^2 - \sum_{i=1}^{p(l)} 1_{\{\xi_i=x\}}(\widehat{\tau_i})^2\right).$$

1. For any function Φ of the discrete loop, show that

$$\mu(\widehat{l}^{x,2}\Phi) = \lambda_x^{-2}\mu\left(\frac{N_x(N_x-1)}{2}1_{\{N_x\ge 2\}}\Phi\right).$$

2. More generally prove in a similar way that

$$\mu(\widehat{l}^{x,(k)}\Phi) = \lambda_x^{-k}\mu\left(\frac{N_x(N_x-1)...(N_x-k+1)}{k!}1_{\{N_x\ge k\}}\Phi\right).$$

Let us come back to the occupation field to compute its Laplace transform. From the Feynman–Kac formula, it comes easily that, denoting $M_{\frac{\chi}{\lambda}}$ the diagonal matrix with coefficients $\frac{\chi_x}{\lambda_x}$

$$\mathbb{P}_t^{x,x}(e^{-\langle\widehat{l},\chi\rangle} - 1) = \frac{1}{\lambda_x}\left(\exp(t(P-I-M_{\frac{\chi}{\lambda}}))_x^x - \exp(t(P-I))_x^x\right).$$

Integrating in t after expanding, we get from the definition of μ (first for χ small enough):

$$\int (e^{-\langle\widehat{l},\chi\rangle} - 1)d\mu(l) = \sum_{k=1}^{\infty}\int_0^{\infty} [Tr((P-M_{\frac{\chi}{\lambda}})^k) - Tr((P)^k)]\frac{t^{k-1}}{k!}e^{-t}dt$$

$$= \sum_{k=1}^{\infty}\frac{1}{k}[Tr((P-M_{\frac{\chi}{\lambda}})^k) - Tr((P)^k)]$$

$$= -Tr(\log(I-P+M_{\frac{\chi}{\lambda}})) + Tr(\log(I-P)).$$

Hence, as $Tr(\log) = \log(\det)$

$$\int (e^{-\langle \hat{l}, \chi \rangle} - 1) d\mu(l) = \log[\det(-L(-L + M_{\chi/\lambda})^{-1})]$$

$$= -\log \det(I + VM_{\hat{\chi}}) = \log \det(I + GM_\chi)$$

which now holds for all non negative χ as both members are analytic in χ. Besides, by the "resolvent" equation (1.1):

$$\det(I + GM_\chi)^{-1} = \det(I - G_\chi M_\chi) = \frac{\det(G_\chi)}{\det(G)}. \qquad (2.13)$$

Note that $\det(I + GM_\chi) = \det(I + M_{\sqrt{\chi}} G M_{\sqrt{\chi}})$ and $\det(I - G_\chi M_\chi) = \det(I - M_{\sqrt{\chi}} G_\chi M_{\sqrt{\chi}})$, so we can deal with symmetric matrices. Finally we have

Proposition 7. $\mu(e^{-\langle \hat{l}, \chi \rangle} - 1) = -\log(\det(I + M_{\sqrt{\chi}} G M_{\sqrt{\chi}})) = \log(\frac{\det(G_\chi)}{\det(G)})$

Note that in particular $\mu(e^{-t\hat{l}^x} - 1) = -\log(1 + tG^{x,x})$. Consequently, the image measure of μ by $\hat{l}^x$ is $1_{\{s>0\}} \frac{1}{s} \exp(-\frac{s}{G^{x,x}}) ds$.

Considering the Laguerre-type polynomials D_k with generating function

$$\sum_1^\infty t^k D_k(u) = e^{\frac{ut}{1+t}} - 1$$

and setting $\sigma_x = G^{x,x}$, we have:

Proposition 8. The variables $\frac{1}{\sqrt{k}} \sigma_x^k D_k(\frac{\hat{l}^x}{\sigma_x})$ are orthonormal in $L^2(\mu)$ for $k > 0$, and more generally

$$\mathbb{E}(\sigma_x^k D_k(\frac{\hat{l}^x}{\sigma_x}) \sigma_y^j D_j(\frac{\hat{l}^y}{\sigma_y})) = \frac{1}{k} \delta_{k,j} (G^{x,y})^{2k}.$$

Proof. By Proposition 7,

$$\int (1 - e^{\frac{\hat{l}^x t}{1+\sigma_x t}})(1 - e^{\frac{\hat{l}^y s}{1+\sigma_y s}}) \mu(dl)$$

$$= \log(1 - \frac{\sigma_x t}{1 + \sigma_x t}) + \log(1 - \frac{\sigma_y s}{1 + \sigma_y s}) - \log \det \begin{pmatrix} 1 - \frac{\sigma_x t}{1+\sigma_x t} & -\frac{tG^{x,y}}{1+\sigma_x t} \\ -\frac{sG^{x,y}}{1+\sigma_y s} & 1 - \frac{\sigma_y s}{1+\sigma_y s} \end{pmatrix}$$

$$= -\log(1 - st(G^{x,y})^2).$$

The proposition follows by expanding both sides in powers of s and t, and identifying the coefficients. $\qquad \square$

Note finally that if χ has support in D, by the restriction property

$$\mu(1_{\{\hat{l}(X\backslash D)=0\}}(e^{-<\hat{l},\chi>}-1)) = -\log(\det(I+M_{\sqrt{\chi}}G^D M_{\sqrt{\chi}})) = \log\Big(\frac{\det(G_\chi^D)}{\det(G^D)}\Big).$$

Here the determinants are taken on matrices indexed by D and G^D denotes the Green function of the process killed on leaving D.

For paths we have $\mathbb{P}_t^{x,y}(e^{-\langle\hat{l},\chi\rangle}) = \frac{1}{\lambda_y}\exp(t(L-M_{\frac{\chi}{\lambda}}))_{x,y}$. Hence

$$\mu^{x,y}(e^{-\langle\hat{\gamma},\chi\rangle}) = \frac{1}{\lambda_y}((I-P+M_{\chi/\lambda})^{-1})_{x,y} = [G_\chi]^{x,y}.$$

In particular, note that from the resolvent equation (1.1), we get that

$$G^{y,x} = [G_{\varepsilon\delta_x}]^{y,x} + \varepsilon[G_{\varepsilon\delta_x}]^{y,x}G^{x,x}.$$

Hence $\frac{[G_{\varepsilon\delta_x}]^{y,x}}{G^{y,x}} = \frac{1}{1+\varepsilon G^{x,x}}$ and therefore, we obtain:

Proposition 9. *Under the probability $\frac{\mu^{y,x}}{G^{y,x}}$, $\hat{l}^x$ follows an exponential distribution of mean $G^{x,x}$.*

Also $\mathbb{E}^x(e^{-\langle\hat{\gamma},\chi\rangle}) = \sum_y[G_\chi]^{x,y}\kappa_y$ i.e. $[G_\chi\kappa]^x$.

Finally, let us note that a direct calculation shows the following result, analogous to Proposition 4 in which the case $x=y$ was left aside.

Proposition 10. *On loops passing through x, $\mu^{x,x}(dl) = \hat{l}^x\mu(dl)$.*

An alternative way to prove the proposition is to check it on multiple local times, using Exercise 11. It can be shown that the algebra formed by linear combinations of multiple local times generates the loop σ-field. Indeed, the discrete loop can be recovered by taking the multiple local time it indexes and noting it is the unique one of maximal index length among non vanishing multiple local times indexed by multiples in which consecutive points are distinct. Then it is easy to get the holding times as the product of any of their powers can be obtained from a multiple local time.

Remark 5. Propositions 4 and 10 can be generalized: For example, if x_i are n points, $\hat{l}^{x_1,\dots,x_n}\mu(dl)$ can be obtained as the image by circular concatenation of the product of the bridge measures $\mu^{x_i,x_{i+1}}(dl)$ and $\prod\hat{l}^{x_i}\mu(dl)$ can be obtained as the sum of the images, by concatenation in all circular orders, of the product of the bridge measures $\mu^{y_{\sigma(i)},x_{\sigma(i+1)}}(dl)$. If (x_i,y_i) are n oriented edges, $\prod\frac{N_{x_i,y_i}(l)}{C_{x_i,y_i}}\mu(dl)$ can be obtained as the sum of the images, by concatenation in all circular orders σ, of the product of the bridge measures $\mu^{y_{\sigma(i)},x_{\sigma(i+1)}}(dl)$. One can also evaluate expressions of the form $\prod\hat{l}^{z_j}\prod\frac{N_{x_i,y_i}(l)}{C_{x_i,y_i}}\mu(dl)$ as a sum of images, by concatenation in all circular orders, of a product of bridge measures .

2.5 Wreath Products

The following construction gives an interesting information about the number of distinct points visited by the loop, which is more difficult to evaluate than the occupation measure.

Associate to each point x of X an integer n_x. Let Z be the product of all the groups $\mathbb{Z}/n_x\mathbb{Z}$. On the wreath product space $X \times Z$, define a set of conductances $\widetilde{C}_{(x,z),(x',z')}$ by:

$$\widetilde{C}_{(x,z),(x',z')} = \frac{1}{n_x n_{x'}} C_{x,x'} \prod_{y \neq x,x'} 1_{\{z_y = z'_y\}}$$

and set $\widetilde{\kappa}_{(x,z)} = \kappa_x$. This means in particular that in the associated Markov chain, the first coordinate is an autonomous Markov chain on X and that in a jump, the Z-configuration can be modified only at the point from which or to which the first coordinate jumps.

Denote by $\widetilde{e}$ the corresponding energy form. Note that $\widetilde{\lambda}_{(x,z)} = \lambda_x$.

Then, denoting $\widetilde{\mu}$ the loop measure and $\widetilde{P}$ the transition matrix on $X \times Z$ defined by $\widetilde{e}$, we have the following

Proposition 11.

$$\prod_{x \in X} n_x \int 1_{\{p>1\}} \prod_{x,\, N_x(l)>0} \frac{1}{n_x} \mu(dl) = \widetilde{\mu}(p > 1) = -\log(\det(I - \widetilde{P})).$$

In particular, if $n_x = n$ for all x,

$$n^{|X|} \int 1_{\{p>1\}} n^{-\#\{x,\, N_x(l)>0\}} \mu(dl) = \widetilde{\mu}(p > 1) = -\log(\det(I - \widetilde{P})).$$

Proof. Each time the Markov chain on $X \times Z$ defined by $\widetilde{e}$ jumps from a point above x to a point above y, z_x and z_y are resampled according to the uniform distribution on $\mathbb{Z}/n_x\mathbb{Z} \times \mathbb{Z}/n_y\mathbb{Z}$, while the other indices z_w are unchanged. It follows that

$$[\widetilde{P}^k]_{(x,z)}^{(x,z)} = \sum_{x_1,\ldots,x_{k-1}} P_{x_1}^x P_{x_2}^{x_1} \ldots P_x^{x_{k-1}} \prod_{y \in \{x,x_1,\ldots,x_{k-1}\}} \frac{1}{n_y}.$$

Note that in the set $\{x, x_1, \ldots, x_{k-1}\}$, distinct points are counted only once, even if the path visit them several times. There are $\prod_{x \in X} n_x$ possible values for z. The detail of the proof is left as an exercise. □

In the case where X is a group and P defines a random walk, $\widetilde{P}$ is associated with a random walk on $X \times Z$ equipped with its wreath product structure (Cf. [38]).

2.6   Countable Spaces

The assumption of finiteness of X can of course be relaxed but we will not do it in detail in these notes, though some infinite examples will be considered. On countable spaces, the previous results can be extended under transience conditions. In this case, the Dirichlet space $\mathbb{H}$ is the space of all functions f with finite energy $e(f)$ which are limits in energy norm of functions with finite support, and the energy defines a Hilbertian scalar product on $\mathbb{H}$.

The energy of a measure is defined as $\sup_{f \in \mathbb{H}} \frac{\mu(f)^2}{e(f)}$. Finitely supported measures have finite energy. Measures of finite energy are elements of the dual $\mathbb{H}^*$ of the Dirichlet space. The potential $G\mu$ is well defined for all finite energy measures μ, by the identity $e(f, G\mu) = \langle f, \mu \rangle$, valid for all f in the Dirichlet space. The energy of the measure μ equals $e(G\mu) = \langle G\mu, \mu \rangle$ (see [10] for more information).

It should also be noted that the submarkovianity of P (i.e. the non negativity of κ) is not essential in the construction of the loop measure μ. It has only to be positive and $I - P$ has to be invertible.

Most important examples of countable graphs are the non ramified covering of finite graphs (Recall that non ramified means that the projection is locally one to one, i.e. that the projection on X of each vertex v of the covering space has the same number of incident edges as v). Consider a non ramified covering graph (Y, F) defined by a normal subgroup H_{x_0} of Γ_{x_0}. The conductances C and the measure λ can be lifted in an obvious way to Y as $H_{x_0} \backslash \Gamma_{x_0}$-periodic functions but the associated Green function $\widehat{G}$ or semigroup are non trivial. By applying $M_\lambda - C$, it is easy to check the following:

Proposition 12. $G^{x,y} = \sum_{\gamma \in H_{x_0} \backslash \Gamma_{x_0}} \widehat{G}^{i(x), \gamma(i(y))}$ *for any section i of the canonical projection from Y onto X.*

Let us consider the universal covering (then H_{x_0} is trivial). It is easy to check it will be transient even in the recurrent case as soon as (X, E) is not circular.

The expression of the Green function $\widehat{G}$ on a universal covering can be given exactly when it is a regular tree, i.e. in the regular graph case. In fact a more general result can be proved as follows:

Given a graph (X, E), set $d_x = \sum_y 1_{\{x,y\} \in E}$ (degree or valency of the vertex x), $D_{x,y} = d_x \delta_{x,y}$ and denote $A_{x,y}$ the incidence matrix $1_E(\{x, y\})$.

Consider the Green function associated with $\lambda_x = (d_x - 1)u + \frac{1}{u}$, with $0 < u < \inf(\frac{1}{d_x - 1}, x \in X)$ and for $\{x, y\} \in E$, $C_{x,y} = 1$.

Proposition 13. *On the universal covering $\mathfrak{T}_{x_0}$, $\widehat{G}^{\,x,y} = u^{d(x,y)} \frac{u}{1 - u^2}$.*

Proof. Note first that as $\frac{1}{u} > d_x - 1$, κ_x is positive for all x. Then $\widehat{G} = (M_\lambda - C)^{-1}$ can be written $\widehat{G} = [u^{-1}I + (D - I)u - A]^{-1}$. Moreover, since we are on a tree,

$$\sum_x A_{z,x} u^{d(x,y)} = (d_z - 1)u^{d(z,y)+1} + u^{d(z,y)-1}$$

for $z \neq y$, hence $\sum_x (\lambda_z \delta_x^z - A_{z,x})u^{d(x,y)} = 0$ for $z \neq y$ and one checks it equals $\frac{1}{u} - u$ for $z = y$. $\qquad\square$

It follows from Proposition 12 that for any section i of the canonical projection from $\mathfrak{T}_{x_0}$ onto X,

$$\sum_{\gamma \in \Gamma_{x_0}} u^{d(i(x),\gamma(i(y)))} = (\frac{1}{u} - u)G^{x,y}.$$

2.7 Zeta Functions for Discrete Loops

We present briefly the terminology of symbolic dynamics (see for example [36]) in this simple framework: Setting $f(x_0, x_1, \ldots, x_n, \ldots) = \log(P_{x_0,x_1})$, P induces the Ruelle operator L_f associated with f.

The pressure is defined as the logarithm of the highest eigenvalue β of P. It is associated with a unique positive eigenfunction h (normalized in $L^2(\lambda)$), by Perron Frobenius theorem. Note that $Ph = \beta h$ implies $\lambda h P = \beta \lambda h$ by duality and that in the recurrent case, the pressure vanishes and $h = \frac{1}{\sqrt{\lambda(X)}}$.

In continuous time, the lowest eigenvalue of $-L$ i.e. $1 - \beta$ plays the role of the pressure.

The equilibrium measure associated with f, $m = h^2\lambda$ is the law of the stationary Markov chain defined by the transition probability $\frac{1}{\beta h_x} P_y^x h_y$.

If $P1 = 1$, i.e. $\kappa = 0$, we can consider a Feynman Kac type perturbation $P^{(\varepsilon\kappa)} = PM_{\frac{\lambda}{\lambda+\varepsilon\kappa}}$, with $\varepsilon \downarrow 0$ and κ a positive measure. Perturbation theory (Cf. for example [13]) shows that $\beta^{(\varepsilon\kappa)} - 1 = \frac{1}{\lambda(X)}\sum_x \frac{\lambda_x}{1+\varepsilon\kappa_x} - 1 + o(\varepsilon) = -\frac{\varepsilon\kappa(X)}{\lambda(X)} + o(\varepsilon)$ and that $h^{(\varepsilon k)} = \frac{1}{\sqrt{\lambda(X)}} + o(\varepsilon)$.

We deduce from that the asymptotic behaviour of

$$\int (e^{-\varepsilon\langle \hat{l}, \chi \rangle} - 1)d\mu^{(\varepsilon\kappa)}(l) = \log(\det(I - P^{(\varepsilon\kappa)})) - \log(\det(I - P^{(\varepsilon(\kappa+\chi))}))$$

which is equivalent to $-\log(1 - \beta^{(\varepsilon(\kappa+\chi))}) + \log(1 - \beta^{(\varepsilon\kappa)})$ and therefore to $\log(\frac{\kappa(X)}{\kappa(X)+\chi(X)})$.

The study of relations between the loop measure μ and the zeta function $(\det(I - sP))^{-1}$ and more generally $(\det(I - M_f P))^{-1}$ with f a function on $[0, 1]$ can be done in the context of discrete loops.

$$\exp\left(\sum_{\substack{\text{based} \\ \text{discrete loops}}} \frac{1}{p(\xi)} s^{p(\xi)} \mu(\xi) \right) = (\det(I - sP))^{-1}$$

can be viewed as a type of zeta function defined for $s \in [0 \ 1/\beta)$

Primitive non trivial (based) discrete loops are defined as discrete based loops which cannot be obtained by the concatenation of $n \geq 2$ identical based loops. Loops are primitive iff they are classes of primitive based loops.

The zeta function has an Euler product expansion: if we denote by ξ° this discrete loop defined by the based discrete loop ξ, and set, for $\xi = (\xi_1, \ldots, \xi_k)$, $\mu(\xi^\circ) = P_{\xi_2}^{\xi_1} P_{\xi_3}^{\xi_2} \ldots P_{\xi_1}^{\xi_k}$, it can be seen, by taking the logarithm, that:

$$(\det(I - sP))^{-1} = \exp\left(\sum_{\substack{\text{based} \\ \text{discrete loops}}} \frac{1}{p(\xi)} s^{p(\xi)} \mu(\xi) \right) = \prod_{\substack{\text{primitive} \\ \text{discrete loops}}} \left(1 - \int s^{p(\xi^\circ)} \mu(\xi^\circ) \right)^{-1}$$

Chapter 3
Geodesic Loops

3.1 Reduction

Given any finite path ω with starting point x_0, the reduced path ω^R is defined as the geodesic arc defined by the endpoint of the lift of ω to $\mathfrak{T}_{x_0}$.

Tree-contour-like based loops can be defined as discrete based loops whose lift to the universal covering are still based loops. Each link is followed the same number of times in opposite directions (backtracking). The reduced path ω^R can equivalently be obtained by removing all tree-contour-like based loops imbedded into it (Fig. 3.1). In particular each loop l based at x_0 defines an element l^R in Γ_{x_0}.

This procedure is an example of *loop erasure*. In any graph, given a path ω, the loop erased path ω^{LE} is defined by removing progressively all based loops imbedded in the path, starting from the origin (Fig. 3.2). It produces a self avoiding path (and we see geodesics in $\mathfrak{T}_{x_0}$ are self avoiding paths). Hence any non ramified covering defines a specific reduction operation by composition of lift, loop erasure, and projection.

3.2 Geodesic Loops and Conjugacy Classes

Then, we can consider loops i.e. equivalence classes of based loops under the natural shift.

Geodesic loops are of particular interest. Note their based loops representatives have to be "tailless": If γ is a geodesic based loop, with $|\gamma| = n$, the tail of γ is defined as $\gamma_1 \gamma_2 ... \gamma_i \gamma_{i-1} ... \gamma_1$ if $i = \sup(j, \gamma_1 \gamma_2 ... \gamma_j = \gamma_{n+1} \gamma_n ... \gamma_{n-j+2})$ with $\gamma_{n+1} = \gamma_1$. The associated geodesic loop is obtained by removing the tail.

The geodesic loops are clearly in bijection with the set of conjugacy classes of the fundamental group. Indeed, if we fix a reference point x_0, a geodesic loop defines the conjugation class formed of the elements of Γ_{x_0} obtained by

Y. Le Jan, *Markov Paths, Loops and Fields*, Lecture Notes in Mathematics 2026, 29
DOI 10.1007/978-3-642-21216-1_3, © Springer-Verlag Berlin Heidelberg 2011

Fig. 3.1 Based loop

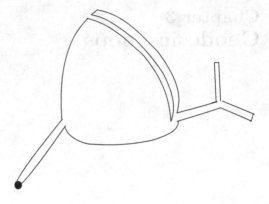

Fig. 3.2 Loop erasure

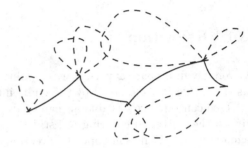

choosing a base point on the loop and a geodesic segment linking it to x_0. Any non trivial element of Γ_{x_0} can be obtained in this way.

Given a loop, there is a canonical geodesic loop associated with it. It is obtained by removing all tails imbedded in it. It can be done by removing one by one all tail edges (i.e. pairs of consecutive inverse oriented edges of the loop). Note that after removal of a tail edge, another tail edge cannot disappear, and that new tail edges appear during this process (Fig. 3.3).

A closed geodesic based at x_0 is called primitive if it cannot be obtained as the concatenation of several identical closed geodesic, i.e. if it is not a non trivial power in Γ_{x_0}. This property is clearly stable under conjugation. Let $\mathfrak{P}$ be corresponding set of primitive geodesic loops. They represent conjugacy classes of primitive elements of Γ (see [50]).

3.3 Geodesics and Boundary

Geodesics lines (half-lines) on a graph are defined as paths without back-tracking indexed by $\mathbb{Z}$ ($\mathbb{N}$).

Paths and in particular geodesics can be defined on (X, E) or on a universal cover $\mathfrak{T}$ and lifted or projected on any intermediate covering space. Two geodesic half lines are said to be confluent if their intersection is a half line.

Fig. 3.3 Loop and
associated geodesic loop

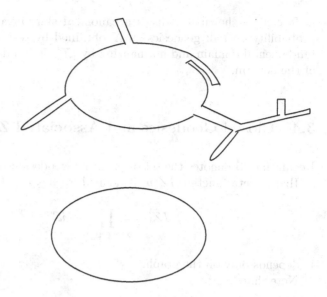

Let us now take the point of view described in Remark 1. Equivalence classes of geodesics half lines of $\mathfrak{T}$ for the confluence relation define the boundary $\partial\mathfrak{T}$ of $\mathfrak{T}$. A geodesic half-line on $\mathfrak{T}$ can therefore be defined by two points: its origin Or and the boundary point θ towards which it converges. It projects on a geodesic half-line on $(X, E) = \Gamma\backslash\mathfrak{T}$. The set of geodesic half lines on (X, E) is identified with $\Gamma\backslash(\mathfrak{T}\times\partial\mathfrak{T})$ which projects canonically onto X.

There is a natural σ-field on the boundary generated by cylinder sets B_g defined by half geodesics starting with a given oriented edge g.

Given any point x_0 in $\mathfrak{T}$, *assuming in this subsection that* $\kappa = 0$, one can define a probability measure on the boundary called the harmonic measure and denoted ν_{x_0}: $\nu_{x_0}(B_g)$ is the probability that the lift of the P-Markov chain starting at x_0 hits g^+ after its last visit to g^-.

Note that Γ acts on the boundary in such a way that $\gamma^*(\nu_x) = \nu_{\gamma(x)}$, for all $\gamma\in\Gamma$ and $x\in\mathfrak{T}$. This harmonic measure induces a probability on the fiber above Γx in $\Gamma\backslash(\mathfrak{T}\times\partial\mathfrak{T})$, i.e. on half geodesics starting at the projection of x on X.

Clearly, in the case of a regular graph, as the universal covering is a r-regular tree, $\nu_{x_0}(B_g) = \frac{1}{r}\frac{1}{(r-1)^{d(x_0,g^-)}}$ where d denotes the distance in the tree. When conductances are all equal, $\nu_{x_0}(B_g)$ can also be computed but is in general distinct from the visibility measure from x_0, $\nu_{x_0}^{vis}(B_g)$, defined as $\frac{1}{d_{x_0}}\prod\frac{1}{d_{x_i}-1}$, $x_1, x_2, ...x_i, ...$ being the points of the geodesic segment linking x_0 to g^-. $\nu_{x_0}^{vis}$ is also a probability on $\partial\mathfrak{T}$.

There is an obvious canonical shift acting on half geodesics.

Note also that $\sum_{x\in\mathfrak{T}}\delta_x^{Or}\nu_x^{vis}(d\theta)$ is a shift-invariant and Γ-invariant measure on the set of half-geodesics of $\mathfrak{T}$.

It can be shown it induces a canonical shift invariant and Γ-invariant probability on half geodesics on X obtained by restricting the sum to any fundamental domain and normalizing by $|X|$. It is independent of the choice of the domain.

3.4 Closed Geodesics and Associated Zeta Function

Recall that $\mathfrak{P}$ denotes the set of primitive geodesic loops.
Ihara's zeta function $IZ(u)$ is defined for $0 \le u < 1$ as

$$IZ(u) = \prod_{\gamma \in \mathfrak{P}} (1 - u^{p(\gamma)})^{-1}$$

It depends only on the graph.
Note that

$$u \frac{\frac{d}{du} IZ(u)}{IZ(u)} = \sum_{\gamma \in \mathfrak{P}} \frac{p(\gamma) u^{p(\gamma)}}{1 - u^{p(\gamma)}} = \sum_{\gamma \in \mathfrak{P}} \sum_{n=1}^{\infty} p(\gamma) u^{np(\gamma)} = \sum_{m=2}^{\infty} N_m u^m$$

where N_m denotes the number of *tailess* geodesic based loops of length m. Indeed, each primitive geodesic loop γ traversed n times still induces $p(\gamma)$ distinct tailess geodesic based loops. Therefore $IZ(u)$ can also be written as $\exp\left(\sum_{m=2}^{\infty} \frac{N_m u^m}{m}\right)$.

Similarly, one can define $\sum_2^{\infty} L_m u^m$ where L_m denotes the number of geodesic based loops of length m. Recall that A denotes the adjacency matrix of the graph, and D the diagonal matrix whose entries are given by the degrees of the vertices.

Assume now that $0 < u < \inf(\frac{1}{d_x - 1}, x \in X)$. We will use again the Green function associated with $\lambda_x = (d_x - 1)u + \frac{1}{u}$ and, for $\{x, y\} \in E$, $C_{x,y} = 1$.

Theorem 1. a) $\sum_2^{\infty} L_m u^m = (1 - u^2) Tr([I + (D - I)u^2 - uA]^{-1}) - |X|$.
(b) $IZ(u) = (1 - u^2)^{-\chi} \det(I - uA + u^2(D - I))^{-1}$ where $\chi =$ denotes the Euler number $|E| - |X|$ of the graph.

Note that an expansion for small u allows to compute the L_m's. In particular, we can check that $L_2 = 0$.

Proof. We adapt the approach of Stark–Terras [50]

(a) As geodesic loops based in x_0 are in bijection with Γ_{x_0}, it follows from Proposition 12 that
$$|V| + \sum_2^{\infty} L_m u^m = (\tfrac{1}{u} - u) Tr(G) = (1 - u^2) Tr([I + (D - I)u^2 - uA]^{-1})$$
(b) Given a geodesic loop l and a base point y of l, let $S_{x,y,l}$ be the sum of the coefficients $u^{p(\delta)}$, where δ varies on all geodesic loops based at x

composed with l and a tail ending at y. If $x = y$, we have $S_{y,y,l} = u^{p(l)}$. Set $S_{x,l} = \sum_y S_{x,y,l}$.

Clearly, for any section i of the canonical projection from $\mathfrak{T}_x$ onto X,

$$\sum_{y,l} S_{x,y,l} = \sum_{\gamma \in \Gamma_x - \{I\}} u^{d(i(x),\gamma i(x))} = (\frac{1}{u} - u)G^{x,x} - 1.$$

On the other hand, considering first the tailless case, then the case where the tail has length 1, and finally decomposing the case where the tail has length at least two according to the position of the point of the tail next to x, (denoted by x'), we obtain the expression:

$$\sum_x S_{x,y,l} = u^{p(l)} + (d_y - 2)u^{p(l)+2} + \sum_{x' \neq y}(d_{x'} - 1)u^2 S_{x',y,l}$$

$$= u^{p(l)} - u^{p(l)+2} + \sum_x (d_x - 1)u^2 S_{x,y,l}$$

summing in y, it comes that

$$\sum_x S_{x,l} = p(l)(u^{p(l)} - u^{p(l)+2}) + \sum_x (d_x - 1)u^2 S_{x,l}.$$

Then, summing on all geodesic loops l

$$\sum_x ((\frac{1}{u} - u)G^{x,x} - 1) = (1-u^2)(\sum N_m u^m) + \sum_x (d_x - 1)u^2((\frac{1}{u} - u)G^{x,x} - 1).$$

Therefore,

$$\sum N_m u^m = Tr((I - u^2(D - I))([I + (D - I)u^2 - uA]^{-1} - (1 - u^2)^{-1}I))$$

$$= Tr(I - (2u^2(D - I) - uA)[I + (D - I)u^2 - uA]^{-1}$$

$$-(1 - u^2)^{-1}(I - u^2(D - I))$$

$$= Tr((2u^2(D - I) - uA)[I + (D - I)u^2 - uA]^{-1} + (1 - u^2)^{-1}(u^2(D - 2I)).$$

To conclude note that

$$\frac{d}{du}\log(\det(I + (D - I)u^2 - uA)) = Tr((2u(D - I) - A)[I + (D - I)u^2 - uA]^{-1})$$

and that $u^2 Tr(D - 2I) = 2u^2 \chi$.

$\square$

An alternative proof

Other proofs can be found in the litterature, especially the following one due
to Kotani–Sunada [14]:

On the line graph, we define a transfer operator Q by $Q^{(y',z)}_{(x,y)} = \delta^{y'}_y 1_{\{z \neq x\}}$.
Then, as $\log(\det(I - uQ)^{-1}) = \sum \frac{u^n}{n} Tr(Q^n)$ and $Tr(Q^n) = N_n$, we have

$$IZ(u) = \det((I - uQ)^{-1}$$

Define the linear map T, from $\mathbb{A}$ to functions on X by $T\alpha(x) = \sum_{y,\{x,y\}\in E} \alpha(x,y)$. Define a linear transformation τ on $\mathbb{A}$ by $\tau\alpha(e) = \alpha(-e)$.
Define S the linear map from functions on X to $\mathbb{A}$ defined by $Sf(x,y) = f(y)$.
Note that $T\tau S = D$, $TS = A$, and $Q = -\tau + ST$.
Then, for any scalar u, $(I - u\tau)(I - uQ) = (1 - u^2)I - (I - u\tau)uST$ and

$$(I - uQ)(I - u\tau) = (1 - u^2)I - uST(I - u\tau). \tag{3.1}$$

Therefore $T(I - u\tau)(I - uQ) = ((1 - u^2)T - uT(I - u\tau)ST) = (I + u^2(D - I) - uA)T$ and

$$T(I - u\tau)(I - uQ)(I - u\tau) = (I + u^2(D - I) - uA)T(I - u\tau).$$

Moreover $(I - uQ)(I - u\tau)S = S((1 - u^2)I - uT(I - u\tau)S)$ and

$$(I - uQ)(I - u\tau)S = S(I + u^2(D - I) - uA). \tag{3.2}$$

It follows from these two last identities that $Im(S)$ and $Ker(T(I - u\tau))$ are
stable under $(I - uQ)(I - u\tau)$.

Note that S is the dual of $-T\tau$: Indeed, for any function f on vertices and
α on oriented edges,

$$\sum_{(x,y)\in E^O} \alpha(x,y)Sf(x,y) = \sum_{(x,y)\in E^O} \alpha(x,y)f(y) = -\sum_y T\tau\alpha(y)f(y).$$

Therefore, $\dim(Im(S)) + \dim(Ker(T\tau)) = 2|E|$.

Note also that $\dim(Ker(T\tau)) = \dim(Ker(T)) = \dim(Ker(T(I - u\tau)))$ (as
$u < 1$).

Moreover, except for a finite set of u's, $Im(S) \cap Ker(T(I - u\tau)) = \{0\}$.
Indeed $T(I - u\tau)S = A - uD$ which is invertible, except for a finite set of u's.

Note that (3.1) implies that $(I - uQ)(I - u\tau)$ equals $(1 - u^2)I$ on $Ker(T(I - u\tau)$ and that (3.2) implies it equals $S(I + u^2(D - I) - uA)S^{-1}$ on $Im(S)$.

It comes that:

$$\det((I - u\tau)(I - uQ)) = (1 - u^2)^{2|E| - |X|} \det(I + u^2(D - I) - uA)$$

On the other hand, $\det((I - u\tau)) = (1 - u^2)^{|E|}$, which allows to conclude.

Chapter 4
Poisson Process of Loops

4.1 Definition

Still following the idea of [18], which was already implicitly in germ in [52], define, for all positive α, the Poissonian ensemble of loops $\mathcal{L}_\alpha$ with intensity $\alpha\mu$.

Note also that these Poissonian ensembles can be considered for fixed α or as a point process of loops indexed by the "time" α. In that case, $\mathcal{L}_\alpha$ is an increasing set of loops with stationary increments. We will denote by $\mathcal{LP}$ the associated Poisson point process of intensity $\mu(dl) \otimes Leb(d\alpha)$ (Leb denoting Lebesgue measure on the positive half-line). It is formed by a countable set of pairs (l_i, α_i) formed by a loop and a time.

We denote by $\mathbb{P}$ its distribution.

Recall that for any functional Φ on the loop space, vanishing on loops of arbitrary small length,

$$\mathbb{E}(e^{i \sum_{l \in \mathcal{L}_\alpha} \Phi(l)}) = \exp(\alpha \int (e^{i\Phi(l)} - 1)\mu(dl))$$

and for any positive functional Ψ on the loops space,

$$\mathbb{E}(e^{-\sum_{l \in \mathcal{L}_\alpha} \Psi(l)}) = \exp(\alpha \int (e^{-\Psi(l)} - 1)\mu(dl)) \tag{4.1}$$

It follows that if Φ is μ-integrable, $\sum_{l \in \mathcal{L}_\alpha} \Phi(l)$ is integrable and

$$\mathbb{E}(\sum_{l \in \mathcal{L}_\alpha} \Phi(l)) = \int \Phi(l)\alpha\mu(dl).$$

Y. Le Jan, *Markov Paths, Loops and Fields*, Lecture Notes in Mathematics 2026, 35
DOI 10.1007/978-3-642-21216-1_4, © Springer-Verlag Berlin Heidelberg 2011

And if in addition Φ^2 is μ-integrable, $\sum_{l \in \mathcal{L}_\alpha} \Phi(l)$ is square-integrable and

$$\mathbb{E}((\sum_{l \in \mathcal{L}_\alpha} \Phi(l)))^2 = \int \Phi^2(l)\alpha\mu(dl) + (\int \Phi(l)\alpha\mu(dl))^2.$$

Recall also "Campbell formula" (Cf. formula 3-13 in [15]): For any system of non negative or μ-integrable loop functionals F_i,

$$\mathbb{E}\Big(\sum_{l_1 \neq l_2 \dots \neq l_k \in \mathcal{L}_\alpha} \prod_1^k F_i(l_i) \Big) = \prod_1^k \alpha\mu(F_i) \tag{4.2}$$

Note the same results hold for functionals of $\mathcal{LP}$.

Of course, $\mathcal{L}_\alpha$ includes trivial loops. The periods τ_l of the trivial loops based at any point x form a Poisson process of intensity $\alpha \frac{e^{-t}}{t}$. It follows directly from this ([37] and references therein) that we have the following.

Proposition 14. *The sum of these periods $\sum \tau_l$ and the set of "frequencies" $\frac{\tau_l}{\sum \tau_l}$ (in decreasing order) are independent and follow respectively a $\Gamma(\alpha)$ and a $Poisson - Dirichlet(0, \alpha)$ distribution.*

Note that by the restriction property, $\mathcal{L}_\alpha^D = \{l \in \mathcal{L}_\alpha, l \subseteq D\}$ is a Poisson process of loops with intensity μ^D, and that $\mathcal{L}_\alpha^D$ is independent of $\mathcal{L}_\alpha \backslash \mathcal{L}_\alpha^D$.

We denote by $\mathcal{DL}_\alpha$ the set of non trivial discrete loops in $\mathcal{L}_\alpha$. Then, $\mathbb{P}(\#\mathcal{DL}_\alpha = k) = e^{-\alpha\mu(p>1)} \frac{\mu(p>1)^k}{k!}$ and conditionally to their number, the discrete loops are independently sampled according to $\frac{1}{\mu(p>1)}\mu 1_{\{p>1\}}$. In particular, if $l_1, l_2, \dots, l_k$ are distinct discrete loops

$$\mathbb{P}(\mathcal{DL}_\alpha = \{l_1, l_2, \dots, l_k\}) = e^{-\alpha\mu(p>1)}\alpha^k \mu(l_1)\dots\mu(l_k)$$

$$= \alpha^k [\frac{\det(G)}{\prod_x \lambda_x}]^\alpha \prod_{x,y} C_{x,y}^{\sum_1^k N_{x,y}(l_i)} \prod_x \lambda_x^{-\sum_1^k N_x(l_i)}.$$

The general result (when the l_i's are not necessarily distinct) follows from the multinomial distibution.

We can associate to $\mathcal{L}_\alpha$ a σ-finite measure (in fact as we will see, it is finite when X is finite, and more generally if G is trace class) called local time or occupation field

$$\widehat{\mathcal{L}_\alpha} = \sum_{l \in \mathcal{L}_\alpha} \hat{l}.$$

Then, for any non-negative measure χ on X

$$\mathbb{E}(e^{-\langle \widehat{\mathcal{L}_\alpha}, \chi \rangle}) = \exp \Big(\alpha \int (e^{-\langle \hat{l}, \chi \rangle} - 1)d\mu(l) \Big).$$

and therefore by Proposition 7 we have

Corollary 1. $\mathbb{E}(e^{-\langle \widehat{\mathcal{L}}_\alpha, \chi \rangle}) = \det(I + M_{\sqrt{\chi}}GM_{\sqrt{\chi}})^{-\alpha} = (\frac{\det(G_\chi)}{\det(G)})^\alpha.$

Many calculations follow from this result.

Note first that $\mathbb{E}(e^{-t\widehat{\mathcal{L}}_\alpha^x}) = (1 + tG^{x,x})^{-\alpha}$. Therefore $\widehat{\mathcal{L}}_\alpha^x$ follows a gamma distribution $\Gamma(\alpha, G^{x,x})$, with density $1_{\{u > 0\}} \frac{e^{-\frac{u}{G^{x,x}}}}{\Gamma(\alpha)} \frac{u^{\alpha-1}}{(G^{xx})^\alpha}$ (in particular, an exponential distribution of mean $G^{x,x}$ for $\alpha = 1$, as $\widehat{l}^x$ under $\frac{\mu^{y,x}}{G^{y,x}}$). When we let α vary as a time parameter, we get a family of gamma subordinators, which can be called a "multivariate gamma subordinator."[1]

We check in particular that $\mathbb{E}(\widehat{\mathcal{L}}_\alpha^x) = \alpha G^{x,x}$ which follows directly from $\mu(\widehat{l}_x) = G^{x,x}.$

Exercise 13. If $\mathcal{L}_\alpha = \{l_i\}$, check that the set of "frequencies" $\frac{\widehat{l}_i^x}{\widehat{\mathcal{L}}_\alpha^x}$ follows a Poisson–Dirichlet distribution of parameters $(0, \alpha)$.

Hint: use the μ-distribution of $\widehat{l}^x$.

Note also that for $\alpha > 1$,

$$\mathbb{E}((1 - \exp(-\frac{\widehat{\mathcal{L}}_\alpha^x}{G^{x,x}}))^{-1}) = \zeta(\alpha).$$

For two points, it follows easily from Corollary 1 that:

$$\mathbb{E}(e^{-t\widehat{\mathcal{L}}_\alpha^x} e^{-s\widehat{\mathcal{L}}_\alpha^y}) = ((1 + tG^{x,x})(1 + sG^{y,y}) - st(G^{x,y})^2)^{-\alpha}$$

This allows to compute the joint density of $\widehat{\mathcal{L}}_\alpha^x$ and $\widehat{\mathcal{L}}_\alpha^y$ in terms of Bessel and Struve functions.

We can condition the loop configuration on the set of associated non trivial discrete loops by using the restricted σ-field $\sigma(\mathcal{DL}_\alpha)$ which contains the variables $N_{x,y}$. We see from (2.11) and (2.7) that

$$\mathbb{E}\Big(e^{-\langle \widehat{\mathcal{L}}_\alpha, \chi \rangle} | \mathcal{DL}_\alpha\Big) = \prod_x (\frac{\lambda_x}{\lambda_x + \chi_x})^{N_x^{(\alpha)} + \alpha}$$

The distribution of $\{N_x^{(\alpha)}, x \in X\}$ follows easily, from Corollary 1 in terms of generating functions:

$$\mathbb{E}(\prod_x s_x^{N_x^{(\alpha)} + 1}) = \det(\delta_{x,y} + \sqrt{\frac{\lambda_x(1 - s_x)}{s_x}} G_{x,y} \sqrt{\frac{\lambda_y(1 - s_y)}{s_y}})^{-\alpha} \qquad (4.3)$$

so that the vector of components $N_x^{(\alpha)}$ follows a multivariate negative binomial distribution (see for example [55]).

---

[1] A subordinator is an increasing Levy process. See for example reference [1].

It follows in particular that $N_x^{(\alpha)}$ follows a negative binomial distribution of parameters $-\alpha$ and $\frac{1}{\lambda_x G^{xx}}$. Note that for $\alpha = 1$, $N_x^{(1)}+1$ follows a geometric distribution of parameter $\frac{1}{\lambda_x G^{xx}}$.

Note finally that in the recurrent case, with the setting and the notations of Sect. 2.7, denoting $\mathcal{L}_{\alpha\varepsilon}^{(\varepsilon\kappa)}$ the Poisson process of loops of intensity $\varepsilon\alpha\mu^{(\varepsilon\kappa)}$, we get that the associated occupation field converges in distribution towards a random constant following a Gamma distribution.

Let us recall one important property of Poisson processes, known as Palm formula.

Proposition 15. *Given any bounded functional Φ on loops configurations and any integrable loop functional F, we have:*

$$\mathbb{E}(\sum_{l\in\mathcal{L}_\alpha} F(l)\Phi(\mathcal{L}_\alpha)) = \int \mathbb{E}(\Phi(\mathcal{L}_\alpha \cup \{l\}))\alpha F(l)\mu(dl).$$

Proof. This is proved by considering first for $\Phi(\mathcal{L}_\alpha)$ the functionals of the form $\sum_{l_1\neq l_2...\neq l_q\in\mathcal{L}_\alpha} \prod_1^q G_j(l_j))$ (with G_j bounded and μ-integrable) which span an algebra separating distinct configurations and applying formula (4.2) : Then, the common value of both members is $\alpha^q \sum_1^q \mu(FG_j) \prod_{l\neq j} \mu(G_l) + \alpha^{q+1}\mu(F) \prod_1^q \mu(G_j)$ □

Exercise 14. Give an alternative proof of this proposition using formula (4.1).

The above proposition applied to $F(l) = \hat{l}^x, N_{x,y}^{(\alpha)}$ and Propositions 4 and 10 yield the following:

Corollary 2.

$$\mathbb{E}(\Phi(\mathcal{L}_\alpha)\widehat{\mathcal{L}_\alpha}^x) = \alpha \int \mathbb{E}(\Phi(\mathcal{L}_\alpha \cup \{\gamma\}))\hat{\gamma}^x\mu(d\gamma) = \alpha \int \mathbb{E}(\Phi(\mathcal{L}_\alpha \cup \{\gamma\}))\mu^{x,x}(d\gamma)$$

and if $x \neq y$

$$\mathbb{E}(\Phi(\mathcal{L}_\alpha)N_{x,y}^{(\alpha)}) = \alpha \int \mathbb{E}(\Phi(\mathcal{L}_\alpha \cup \{\gamma\}))N_{x,y}(\gamma)\mu(d\gamma)$$

$$= \alpha C_{x,y} \int \mathbb{E}(\Phi(\mathcal{L}_\alpha \cup \{\gamma\}))\mu^{x,y}(d\gamma).$$

Remark 6. Proposition 15 and Corollary 2 can be easily generalized to functionals of the Poisson process $\mathcal{L}P$.

Exercise 15. Generalize Corollary 2 to $\widehat{\mathcal{L}_\alpha}^x\widehat{\mathcal{L}_\alpha}^y$, for $x \neq y$.

4.2 Moments and Polynomials of the Occupation Field

It is easy to check (and well known from the properties of the gamma distributions) that the moments of $\widehat{\mathcal{L}_\alpha}^x$ are related to the factorial moments of $N_x^{(\alpha)}$:

$$\mathbb{E}((\widehat{\mathcal{L}_\alpha}^x)^k | \mathcal{DL}_\alpha) = \frac{(N_x^{(\alpha)} + k - 1 + \alpha)(N_x^{(\alpha)} + k - 2 + \alpha)...(N_x^{(\alpha)} + \alpha)}{\lambda_x^k}$$

Exercise 16. Denoting $\mathcal{L}_\alpha^+$ the set of non trivial loops in $\mathcal{L}_\alpha$, define

$$\widehat{\mathcal{L}_\alpha}^{x,(k)} = \sum_{m=1}^{k} \sum_{k_1+...+k_m=k} \sum_{l_1 \neq l_2 ... \neq l_m \in \mathcal{L}_\alpha^+} \prod_{j=1}^{m} \widehat{l_j}^{x,(k_j)}.$$

Deduce from Exercise 12 that $\mathbb{E}(\widehat{\mathcal{L}_\alpha}^{x,(k)} | \mathcal{DL}_\alpha) = \frac{1}{\lambda_x^k} 1_{\{N_x \geq k\}} (N_x^{(\alpha)} - k + 1)...(N_x^{(\alpha)} - 1)N_x^{(\alpha)}$

It is well known that Laguerre polynomials $L_k^{(\alpha-1)}$ with generating function

$$\sum_{0}^{\infty} t^k L_k^{(\alpha-1)}(u) = \frac{e^{-\frac{ut}{1-t}}}{(1-t)^\alpha}$$

are orthogonal for the $\Gamma(\alpha)$ distribution. They have mean zero and variance $\frac{\Gamma(\alpha+k)}{k!\Gamma(\alpha)}$. Hence if we set $\sigma_x = G^{x,x}$ and $P_k^{\alpha,\sigma}(x) = (-\sigma)^k L_k^{(\alpha-1)}(\frac{x}{\sigma})$, the random variables $P_k^{\alpha,\sigma_x}(\widehat{\mathcal{L}_\alpha}^x)$ are orthogonal with mean 0 and variance $\sigma^{2k} \frac{\Gamma(\alpha+k)}{k!\Gamma(\alpha)}$, for $k > 0$.

Note that $P_1^{\alpha,\sigma_x}(\widehat{\mathcal{L}_\alpha}^x) = \widehat{\mathcal{L}_\alpha}^x - \alpha\sigma_x = \widehat{\mathcal{L}_\alpha}^x - \mathbb{E}(\widehat{\mathcal{L}_\alpha}^x)$. It will be denoted $\widetilde{\mathcal{L}_\alpha}^x$.

Moreover, we have $\sum_0^\infty t^k P_k^{\alpha,\sigma}(u) = \sum(-\sigma t)^k L_k^{(\alpha-1)}(\frac{u}{\sigma}) = \frac{e^{\frac{ut}{1+\sigma t}}}{(1+\sigma t)^\alpha}$

Note that by Corollary 1,

$$\mathbb{E}\left(\frac{e^{\frac{\widehat{\mathcal{L}_\alpha}^x t}{1+\sigma_x t}}}{(1+\sigma_x t)^\alpha} \frac{e^{\frac{\widehat{\mathcal{L}_\alpha}^y s}{1+\sigma_y s}}}{(1+\sigma_y s)^\alpha} \right)$$

$$= \frac{1}{(1+\sigma_x t)^\alpha (1+\sigma_y s)^\alpha} ((1 - \frac{\sigma_x t}{1+\sigma_x t})(1 - \frac{\sigma_y s}{1+\sigma_y s}) - \frac{t}{1+\sigma_x t}\frac{s}{1+\sigma_y s}(G^{x,y})^2)^{-\alpha}$$

$$= (1 - st(G^{x,y})^2)^{-\alpha}.$$

Therefore, we get, by developing in entire series in (s, t) and identifying the coefficients:

$$\mathbb{E}(P_k^{\alpha, \sigma_x}(\widehat{\mathcal{L}_\alpha}^x), P_l^{\alpha, \sigma_y}(\widehat{\mathcal{L}_\alpha}^y)) = \delta_{k,l}(G^{x,y})^{2k} \frac{\alpha(\alpha + 1)...(\alpha + k - 1)}{k!} \quad (4.4)$$

Let us stress the fact that $G^{x,x}$ and $G^{y,y}$ do not appear on the right hand side of this formula. This is quite important from the renormalisation point of view, as we will consider in the last section the two dimensional Brownian motion for which the Green function diverges on the diagonal.

More generally one can prove similar formulas for products of higher order.

It should also be noted that if we let α increase, $(1 + \sigma_x t)^{-\alpha} \exp(\frac{\widehat{\mathcal{L}_\alpha}^x t}{1 + \sigma_x t})$ and $P_k^{\alpha, \sigma_x}(\widehat{\mathcal{L}_\alpha}^x)$ are $\sigma(\mathcal{L}_\alpha)$-martingales with expectations respectively equal to 1 and 0.

Note that since $G_\chi M_\chi$ is a contraction, from determinant expansions given in [54] and [55], we have

$$\det(I + M_{\sqrt{\chi}} G M_{\sqrt{\chi}})^{-\alpha} = 1 + \sum_{k=1}^{\infty} \frac{(-1)^k}{k!} \sum \chi_{i_1}...\chi_{i_k} Per_\alpha(G_{i_l, i_m}, 1 \leq l, m \leq k).$$

(4.5)

The α-permanent Per_α is defined as $\sum_{\sigma \in S_k} \alpha^{m(\sigma)} G_{i_1, i_{\sigma(1)}}...G_{i_k, i_{\sigma(k)}}$ with $m(\sigma)$ denoting the number of cycles in σ. Then, from Corollary 1, it follows that:

$$\mathbb{E}(\langle \widehat{\mathcal{L}_\alpha}, \chi \rangle^k) = \sum \chi_{i_1}...\chi_{i_k} Per_\alpha(G_{i_l, i_m}, 1 \leq l, m \leq k).$$

Note that an explicit form for the multivariate negative binomial distribution, and therefore, a series expansion for the density of the multivariate gamma distribution, follows directly (see [55]) from this determinant expansion.

It is actually not difficult to give a direct proof of this result. Thus, the Poisson process of loops provides a natural probabilistic proof and interpretation of this combinatorial identity (see [55] for an historical view of the subject).

We can show in fact that:

Proposition 16. *For any $(x_1, ...x_k)$ in X^k, $\mathbb{E}(\widehat{\mathcal{L}_\alpha}^{x_1}...\widehat{\mathcal{L}_\alpha}^{x_k}) = Per_\alpha(G^{x_l, x_m}, 1 \leq l, m \leq k)$*

Proof. The cycles of the permutations in the expression of Per_α are associated with point configurations on loops. We obtain the result by summing the contributions of all possible partitions of the points $i_1...i_k$ into a finite set of distinct loops. We can then decompose again the expression according to ordering of points on each loop. We can conclude by using the formula $\mu(l^{x_1, ..., x_m}) = G^{x_1, x_2} G^{x_2, x_3}...G^{x_m, x_1}$ and Campbell formula (4.2). $\qquad \square$

Remark 7. We can actually, in the special case $i_1 = i_2 = \ldots = i_k = x$, check this formula in a different way. From the moments of the Gamma distribution, we have that $\mathbb{E}((\widehat{\mathcal{L}_\alpha}^x)^n) = (G^{x,x})^n \alpha(\alpha+1)\ldots(\alpha+n-1)$ and the α-permanent can be written $\sum_1^n d(n,k)\alpha^k$ where the coefficients $d(n,k)$ are the numbers of n-permutations with k cycles (Stirling numbers of the first kind). One checks that $d(n+1,k) = nd(n,k) + d(n,k-1)$.

Let $\mathcal{S}_k^0$ be the set of permutations of k elements without fixed point. They correspond to configurations without isolated points.

Set $Per_\alpha^0(G^{i_l,i_m}, 1 \le l,m \le k) = \sum_{\sigma \in \mathcal{S}_k^0} \alpha^{m(\sigma)} G^{i_1,i_{\sigma(1)}}\ldots G^{i_k,i_{\sigma(k)}}$. Then an easy calculation shows that:

Corollary 3. $\mathbb{E}(\widetilde{\mathcal{L}_\alpha}^{i_1}\ldots\widetilde{\mathcal{L}_\alpha}^{i_k}) = Per_\alpha^0(G^{i_l,i_m}, 1 \le l,m \le k)$

Proof. Indeed, the expectation can be written

$$\sum_{p \le k} \sum_{I \subseteq \{1,\ldots k\}, |I|=p} (-1)^{k-p} \prod_{l \in I^c} G^{i_l,i_l} Per_\alpha(G^{i_a,i_b}, a,b \in I)$$

and

$$Per_\alpha(G^{i_a,i_b}, a,b \in I) = \sum_{J \subseteq I} \prod_{j \in I \setminus J} G^{j,j} Per_\alpha^0(G^{i_a,i_b}, a,b \in J).$$

Then, expressing $\mathbb{E}(\widetilde{\mathcal{L}_\alpha}^{i_1}\ldots\widetilde{\mathcal{L}_\alpha}^{i_k})$ in terms of Per_α^0's, we see that if $J \subseteq \{1,\ldots k\}$, $|J| < k$, the coefficient of $Per_\alpha^0(G^{i_a,i_b}, a,b \in J)$ is $\sum_{I, I \supseteq J}(-1)^{k-|I|} \prod_{j \in J^c} G^{i_j,i_j}$ which vanishes as $(-1)^{-|I|} = (-1)^{|I|} = (-1)^{|J|}(-1)^{|I \setminus J|}$ and $\sum_{I \supseteq J}(-1)^{|I \setminus J|} = (1-1)^{k-|J|} = 0$. $\square$

Set $Q_k^{\alpha,\sigma}(u) = P_k^{\alpha,\sigma}(u + \alpha\sigma)$ so that $P_k^{\alpha,\sigma}(\widehat{\mathcal{L}_\alpha}^x) = Q_k^{\alpha,\sigma}(\widehat{\mathcal{L}_\alpha}^x)$. This quantity will be called the n-th renormalized self intersection local time or the n-th renormalized power of the occupation field and denoted $\widetilde{\mathcal{L}}_\alpha^{x,n}$.

From the recurrence relation of Laguerre polynomials

$$nL_n^{(\alpha-1)}(u) = (-u + 2n + \alpha - 2)L_{n-1}^{(\alpha-1)} - (n + \alpha - 2)L_{n-2}^{(\alpha-1)},$$

we get that

$$nQ_n^{\alpha,\sigma}(u) = (u - 2\sigma(n-1))Q_{n-1}^{\alpha,\sigma}(u) - \sigma^2(\alpha + n - 2)Q_{n-2}^{\alpha,\sigma}(u).$$

In particular $Q_2^{\alpha,\sigma}(u) = \frac{1}{2}(u^2 - 2\sigma u - \alpha\sigma^2)$, $Q_3^{\alpha,\sigma}(u) = \frac{1}{6}(u^3 - 6\sigma u^2 + 3u\sigma^2(2-\alpha) + 4\sigma^3\alpha)$.

We have also, from (4.4)

$$\mathbb{E}(Q_k^{\alpha,\sigma_x}(\widehat{\mathcal{L}_\alpha}^x), Q_l^{\alpha,\sigma_y}(\widehat{\mathcal{L}_\alpha}^y)) = \delta_{k,l}(G^{x,y})^{2k}\frac{\alpha(\alpha+1)\ldots(\alpha+k-1)}{k!} \qquad (4.6)$$

The comparison of the identity (4.6) and Corollary 3 yields a combinatorial result which will be extended in the renormalizing procedure presented in the last section.

The identity (4.6) can be considered as a polynomial identity in the variables σ_x, σ_y and $G^{x,y}$.

Set $Q_k^{\alpha,\sigma_x}(u) = \sum_{m=0}^{k} q_m^{\alpha,k} u^m \sigma_x^{k-m}$, and denote $N_{n,m,r,p}$ the number of ordered configurations of n black points and m red points on r non trivial oriented cycles, such that only $2p$ links are between red and black points. We have first by Corollary 3:

$$\mathbb{E}((\widetilde{\mathcal{L}_\alpha}^x)^n (\widetilde{\mathcal{L}_\alpha}^y)^m) = \sum_{r} \sum_{p \leq \inf(m,n)} \alpha^r N_{n,m,r,p} (G^{x,y})^{2p} (\sigma_x)^{n-p} (\sigma_y)^{m-p}$$

and therefore

$$\sum_{r} \sum_{p \leq m \leq k} \sum_{p \leq n \leq l} \alpha^r q_m^{\alpha,k} q_n^{\alpha,l} N_{n,m,r,p} = 0 \text{ unless } p = l = k. \tag{4.7}$$

$$\sum_{r} \alpha^r q_k^{\alpha,k} q_k^{\alpha,k} N_{k,k,r,k} = \frac{\alpha(\alpha+1)...(\alpha+k-1)}{k!}. \tag{4.8}$$

Note that one can check directly that $q_k^{\alpha,k} = \frac{1}{k!}$, and $N_{k,k,1,k} = k!(k-1)!$, $N_{k,k,k,k} = k!$ which confirms the identity (4.8) above.

Exercise 17. Let $d_{n,k}^0$ be the number of n-permutations with no fixed points and k cycles. If k_j, $2 \leq j \leq n$ are integers such that $\sum_j j k_j = m$ and $\sum_j k_j = k$, let $C_{m,k}(k_j, 2 \leq j \leq n)$ be the number of m-permutations with no fixed points and k_j cycles of length j. Note that $\sum C_{m,k}(k_j, 2 \leq j \leq n) = d_{m,k}^0$. Show the following identities:

(a) $\sum_0^n \binom{n}{m} d_{m,k}^0 = d_{n,k}$ (the number of n-permutations with k cycles)

(b) $C_{m,k}(k_j, 2 \leq j \leq n) = \frac{m!}{\prod k_j! j^{k_j}}$

(c) $\sum t^N Q_N^{\alpha,\sigma}(u) = e^{\frac{t(u+\alpha\sigma)}{1+t\sigma}} (1 - \frac{t\sigma}{1+t\sigma})^\alpha = e^{\frac{tu}{1+t\sigma}} (1 + \sum_{m=1}^{\infty} \sum_{k=1}^{m} d_{m,k}^0 \frac{(-\alpha)^k}{m!} (\frac{t\sigma}{1+t\sigma})^m)$ (Hint: Use Corollaries 1 and 3 to prove this equality)

$= \sum_{l=0}^{\infty} \frac{t^l u^l}{l!} (1+t\sigma)^{-l} + \sum_{l=0}^{\infty} \sum_{m=1}^{\infty} \frac{t^{l+m} u^l}{m! l!} (1+t\sigma)^{-m-l} \sum_{1 \leq k \leq m} d_{m,k}^0 (-\alpha)^k$

(d) $Q_N^{\alpha,\sigma}(u) = \sum_{0 \leq l \leq N} \sum_{k \leq N-l} a_{N,l,k} u^l \sigma^{N-l} \alpha^k$

with $a_{N,l,k} = \sum_{m=k}^{N-l} (-1)^{N-l-k-m} \frac{(N-1)!}{l! m! (N-l-m)! (m+l-1)!} d_{m,k}^0$ for $k \geq 1$,

$a_{N,l,0} = (-1)^{N-l-k} \frac{(N-1)!}{l! (N-l)! (l-1)!}$ and $a_{N,0,0} = 0$

4.3 Hitting Probabilities

Let us start with some elements of discrete potential theory. Denote by

$$[H^F]^x_y = \mathbb{P}_x(x_{T_F} = y)$$

the hitting distribution of F by the Markov chain starting at x (H^F is called the *balayage or Poisson kernel* in Potential theory). Set $D = F^c$ and denote by e^D, $P^D = P|_{D \times D}$, $V^D = [(I - P^D)]^{-1}$ and $G^D = [(M_\lambda - C)|_{D \times D}]^{-1}$ the energy, the transition matrix, the potential and the Green function of the Markov chain killed at the hitting time of F.

Denote by $\mathbb{P}^D_x$ the law of the killed Markov chain starting at x.

Hitting probabilities can be expressed in terms of Green functions. For $y \in F$, we have

$$[H^F]^x_y = 1_{\{x=y\}} + \sum_0^\infty \sum_{z \in D} [(P^D)^k]^x_z P^z_y$$

We see also that
$$V = V^D + H^F V$$

and that
$$G = G^D + H^F G$$

As G and G^D are symmetric, we have $[H^F G]^x_y = [H^F G]^y_x$ so that for any measure ν,
$$H^F(G\nu) = G(\nu H^F).$$

In particular, the *capacitary potential* $H^F 1$ is the potential of the *capacitary measure* κH^F.

Therefore we see that for any function f and measure ν,

$$e(H^F f, G^D \nu) = e(H^F f, G\nu) - e(H^F f, H^F G\nu) = \left\langle H^F f, \nu \right\rangle - e(H^F f, G(H^F \nu)) = 0$$

as $(H^F)^2 = H^F$.

Equivalently, we have the following:

Proposition 17. *For any g vanishing on F, $e(H^F f, g) = 0$ so that $I - H^F$ is the e-orthogonal projection on the space of functions supported in D.*

The energy of the capacitary potential of F, $e(H^F 1, H^F 1)$ equals the mass of the capacitary measure $\left\langle \kappa H^F, 1 \right\rangle$. It is called the *capacity* of F and denoted $Cap_e(F)$.

Note that some of these results extend without difficulty to the recurrent case. In particular, for any measure ν supported in D, $G^D \nu = G(\nu - \nu H^F)$

and $e(H^F f, G^D \nu) = 0$ for all f. For further developments see for example [22] and its references.

The restriction property holds for $\mathcal{L}_\alpha$ as it holds for μ. The set $\mathcal{L}_\alpha^D$ of loops inside D is associated with μ^D and is independent of $\mathcal{L}_\alpha - \mathcal{L}_\alpha^D$. Therefore, we see from Corollary 1 that

$$\mathbb{E}(e^{-\langle \widehat{\mathcal{L}_\alpha - \mathcal{L}_\alpha^D}, x \rangle}) = \left(\frac{\det(G_\chi)}{\det(G)} \frac{\det(G^D)}{\det(G_\chi^D)} \right)^\alpha.$$

Note that for all x, $\mu(\widehat{l^x} > 0) = \infty$. This is due to trivial loops and it can be seen directly from the definition of μ that in this simple framework the loops of $\mathcal{L}_\alpha$ cover the whole space X.

Note however that

$$\mu(\widehat{l}(F) > 0, p > 1) = \mu(p > 1) - \mu(\widehat{l}(F) = 0, p > 1) = \mu(p > 1) - \mu^D(p > 1)$$

$$= -\log(\frac{\det(I - P)}{\det_{D \times D}(I - P)}) = -\log(\frac{\det(G^D)}{\prod_{x \in F} \lambda_x \det(G)}).$$

It follows that the probability that no non trivial loop (i.e. a loop which is not reduced to a point) in $\mathcal{L}_\alpha$ intersects F equals

$$\exp(-\alpha \mu(\{l, p(l) > 1, \widehat{l}(F) > 0\})) = \left(\frac{\det(G^D)}{\prod_{x \in F} \lambda_x \det(G)} \right)^\alpha.$$

Recall Jacobi's identity: for any $(n+p, n+p)$ invertible matrix A, denoting e_i the canonical basis,

$$\det(A^{-1}) \det(A_{ij}, 1 \le i, j \le n) = \det(A^{-1}) \det(Ae_1, ..., Ae_n, e_{n+1}, ..., e_{n+p})$$

$$= \det(e_1, ..., e_n, A^{-1}e_{n+1}, ..., A^{-1}e_{n+p})$$

$$= \det((A^{-1})_{k,l}, n \le k, l \le n + p).$$

In particular, $\det(G^D) = \frac{\det(G)}{\det(G|_{F \times F})}$, we can also denote $\frac{\det(G)}{\det_{F \times F}(G)}$. So we have the

Proposition 18. *The probability that no non-trivial loop in $\mathcal{L}_\alpha$ intersects F equals*

$$[\prod_{x \in F} \lambda_x \det_{F \times F}(G)]^{-\alpha}.$$

Moreover $\mathbb{E}(e^{-\langle \widehat{\mathcal{L}_\alpha - \mathcal{L}_\alpha^D}, x \rangle}) = (\frac{\det_{F \times F}(G_\chi)}{\det_{F \times F}(G)})^\alpha.$

In particular, it follows that the probability that no non-trivial loop in $\mathcal{L}_\alpha$ visits x equals $(\frac{1}{\lambda_x G^{x,x}})^\alpha$ which is also a consequence of the fact that N_x follows a negative binomial distribution of parameters $-\alpha$ and $\frac{1}{\lambda_x G^{x,x}}$.

Also, if F_1 and F_2 are disjoint,

$$\mu(\widehat{l}(F_1)\widehat{l}(F_2)) > 0) = \mu(\widehat{l}(F_1) > 0, p > 1) + \mu(\widehat{l}(F_2) > 0, p > 1)$$

$$- \mu(\widehat{l}(F_1 \cup F_2) > 0, p > 1)$$

$$= \log(\frac{\det(G)\det(G^{D_1 \cap D_2})}{\det(G^{D_1})\det(G^{D_2})})$$

$$= \log(\frac{\det_{F_1 \times F_1}(G)\det_{F_2 \times F_2}(G)}{\det_{F_1 \cup F_2 \times F_1 \cup F_2}(G)}). \tag{4.9}$$

Therefore the probability that no loop in $\mathcal{L}_\alpha$ intersects F_1 and F_2 equals

$$\exp(-\alpha\mu(\{l, \prod \widehat{l}(F_i) > 0\})) = (\frac{\det(G^{D_1})\det(G^{D_2})}{\det(G)\det(G^{D_1 \cap D_2})})^\alpha$$

$$= (\frac{\det_{F_1 \times F_1}(G)\det_{F_2 \times F_2}(G)}{\det_{F_1 \cup F_2 \times F_1 \cup F_2}(G)})^{-\alpha}$$

It follows that the probability no loop in $\mathcal{L}_\alpha$ visits two distinct points x and y equals $(\frac{G^{x,x}G^{y,y}-(G^{x,y})^2}{G^{x,x}G^{y,y}})^\alpha$ and in particular $1 - \frac{(G^{x,y})^2}{G^{x,x}G^{y,y}}$ if $\alpha = 1$.

Exercise 18. Generalize this formula to n disjoint sets:

$$\mathbb{P}(\nexists l \in \mathcal{L}_\alpha, \prod \widehat{l}(F_i) > 0) = \left(\frac{\det(G)\prod_{i<j}\det(G^{D_i \cap D_j})...}{\prod \det(G^{D_i})\prod_{i<j<k}\det(G^{D_i \cap D_j \cap D_k})...}\right)^{-\alpha}$$

Note this yields an interesting determinant product inequality.

Chapter 5
The Gaussian Free Field

5.1 Dynkin's Isomorphism

By a well known calculation on Gaussian measure, if X is finite, for any $\chi \in \mathbb{R}_+^X$,

$$\frac{\sqrt{\det(M_\lambda - C)}}{(2\pi)^{|X|/2}} \int_{\mathbb{R}^X} e^{-\frac{1}{2}\sum \chi_u (v^u)^2} e^{-\frac{1}{2}e(v)} \Pi_{u \in X} dv^u = \sqrt{\frac{\det(G_\chi)}{\det(G)}}$$

and

$$\frac{\sqrt{\det(M_\lambda - C)}}{(2\pi)^{|X|/2}} \int_{\mathbb{R}^X} v^x v^y e^{-\frac{1}{2}\sum \chi_u (v^u)^2} e^{-\frac{1}{2}e(v)} \Pi_{u \in X} dv^u = (G_\chi)^{x,y} \sqrt{\frac{\det(G_\chi)}{\det(G)}}$$

This can be easily reformulated by introducing on an independent probability space the Gaussian free field ϕ defined by the covariance $\mathbb{E}_\phi(\phi^x \phi^y) = G^{x,y}$ (this reformulation cannot be dispensed when X becomes infinite)
So we have

$$\mathbb{E}_\phi(e^{-\frac{1}{2}<\phi^2,\chi>}) = \det(I + GM_\chi)^{-\frac{1}{2}} = \sqrt{\det(G_\chi G^{-1})}$$

and

$$\mathbb{E}_\phi(\phi^x \phi^y e^{-\frac{1}{2}<\phi^2,\chi>}) = (G_\chi)^{x,y} \sqrt{\det(G_\chi G^{-1})}.$$

Then since sums of exponentials of the form $e^{-\frac{1}{2}<\cdot,\chi>}$ are dense in continuous functions on $\mathbb{R}_+^X$ the following holds:

Theorem 2. (a) The fields $\widehat{\mathcal{L}}_{\frac{1}{2}}$ and $\frac{1}{2}\phi^2$ have the same distribution.

(b) $\mathbb{E}_\phi((\phi^x \phi^y F(\frac{1}{2}\phi^2)) = \int \mathbb{E}(F(\widehat{\mathcal{L}}_{\frac{1}{2}} + \widehat{\gamma}))\mu^{x,y}(d\gamma)$ for any bounded functional F of a non negative field.

Y. Le Jan, *Markov Paths, Loops and Fields*, Lecture Notes in Mathematics 2026, 47
DOI 10.1007/978-3-642-21216-1_5, © Springer-Verlag Berlin Heidelberg 2011

Remarks:

(a) This can be viewed as a version of Dynkin's isomorphism (Cf. [6]). It can be extended to non-symmetric generators (Cf. [24]).

(b) For $x = y$, (b) follows from (a) and Corollary 2. Also, by the same corollary if $C_{x,y} \neq 0$, (b) implies that

$$\mathbb{E}_\phi(\phi^x \phi^y F(\frac{1}{2}\phi^2)) = \frac{2}{C_{x,y}}\mathbb{E}(F(\widehat{\mathcal{L}_{\frac{1}{2}}})N_{x,y}^{(\frac{1}{2})})$$

(c) An analogous result can be given when α is any positive half integer, by using real vector valued Gaussian field, or equivalently complex fields for integral values of α (in particular $\alpha = 1$): If $\vec{\phi} = (\phi_1, \phi_2, ..., \phi_k)$ are k independent copies of the real free field, the fields $\widehat{\mathcal{L}_{\frac{k}{2}}}$ and $\frac{1}{2}\left\|\vec{\phi}\right\|^2 = \frac{1}{2}\sum_1^k \phi_j^2$ have the same law and

$$\mathbb{E}_{\vec{\phi}}(\left\langle \vec{\phi}^x, \vec{\phi}^y \right\rangle F(\frac{1}{2}\|\phi\|^2)) = k \int \mathbb{E}(F(\widehat{\mathcal{L}_{\frac{k}{2}}} + \widehat{\gamma}))\mu^{x,y}(d\gamma).$$

The complex free field $\phi_1 + i\phi_2$ will be denoted φ. If we consider k independent copies φ_j of this field, $\widehat{\mathcal{L}_k}$ and $\frac{1}{2}\|\vec{\varphi}\|^2 = \frac{1}{2}\sum_1^k \varphi_j\overline{\varphi_j}$ have the same law.

(d) Note it implies immediately that the process ϕ^2 is infinitely divisible. See [9] and its references for a converse and earlier proofs of this last fact.

Theorem 2 suggests the following:

Exercise 19. Show that for any bounded functional F of a non negative field, if x_i are $2k$ points:

$$\mathbb{E}_\phi(F(\phi^2)\prod \phi^{x_i}) = \int \mathbb{E}(F(\widehat{\mathcal{L}_{\frac{1}{2}}} + \sum_1^k \widehat{\gamma_j})) \sum_{\text{pairings}} \prod_{pairs} \mu^{y_j,z_j}(d\gamma_j)$$

where $\sum_{\text{pairings}}$ means that the k pairs y_j, z_j are formed with all the $2k$ points x_i, in all $\frac{(2k)!}{2^k k!}$ possible ways.

Hint: As in the proof of Theorem 2, we take F of the form $e^{-\frac{1}{2}<\cdot,\chi>}$. Then we use the classical expression for the expectation of a product of Gaussian variables known as Wick theorem (see for example [35, 48]).

Exercise 20. For any f in the Dirichlet space $\mathbb{H}$ of functions of finite energy (i.e. for all functions if X is finite), the law of $f + \phi$ is absolutely continuous with respect to the law of ϕ, with density $\exp(< -Lf, \phi >_m -\frac{1}{2}e(f))$.

Exercise 21. (a) Using Proposition 17, show (it was observed by Nelson in the context of the classical (or Brownian) free field) that the Gaussian field ϕ is Markovian: Given any subset F of X, denote $\mathcal{H}_F$ the Gaussian space spanned by $\{\phi^y, y \in F\}$. Then, for $x \in D = F^c$, the projection of ϕ^x on $\mathcal{H}_F$ (i.e. the conditional expectation of ϕ^x given $\sigma(\phi^y, y \in F)$) is $\sum_{y \in F} [H^F]_y^x \phi^y$.

(b) Moreover, show that $\phi^D = \phi - H^F \phi$ is the Gaussian free field associated with the process killed at the exit of D.

5.2 Wick Products

We have seen in Theorem 2 that L^2 functionals of $\widehat{\mathcal{L}_1}$ can be represented in this space of Gaussian functionals. In order to prepare the extension of this representation to the more difficult framework of continuous spaces (which can often be viewed as scaling limits of discrete spaces), including especially the planar Brownian motion considered in [18], we shall introduce the renormalized (or Wick) powers of ϕ. We set : $(\phi^x)^n := (G^{x,x})^{\frac{n}{2}} H_n(\phi^x / \sqrt{G^{x,x}})$ where H_n in the n-th Hermite polynomial (characterized by $\sum \frac{t^n}{n!} H_n(u) = e^{tu - \frac{t^2}{2}}$). These variables are orthogonal in L^2 and $E((: (\phi^x)^n :)^2) = n! G^{x,x}$.

Setting as before $\sigma_x = G^{x,x}$, from the relation between Hermite polynomials H_{2n} and Laguerre polynomials $L_n^{-\frac{1}{2}}$,

$$H_{2n}(x) = (-2)^n n! L_n^{-\frac{1}{2}}(\frac{x^2}{2})$$

it follows that:

$$: (\phi^x)^{2n} := 2^n n! P_n^{\frac{1}{2}, \sigma}((\frac{(\phi^x)^2}{2})).$$

and

$$\mathbb{E}_\phi((: (\phi^x)^n :)^2) = \sigma_x^n n!$$

More generally, if $\phi_1, \phi_2, ..., \phi_k$ are k independent copies of the free field, we can define: $\prod_{j=1}^k (\phi_j^x)^{n_j} := \prod_{j=1}^k : (\phi_j^x)^{n_j} :$. Then it follows that:

$$: (\sum_1^k ((\phi_j^x)^2))^n := \sum_{n_1 + .. + n_k = n} \frac{n!}{n_1! ... n_k!} \prod_{j=1}^k : (\phi_j^x)^{2n_j} :.$$

On the other hand, from the generating function of the polynomials $P_n^{\frac{k}{2}, \sigma}$, we get easily that

$$P_n^{\frac{k}{2}, \sigma}(\sum_1^k u_j) = \sum_{n_1 + .. + n_k = n} \prod_{j=1}^k P_{n_j}^{\frac{1}{2}, \sigma}(u_j).$$

Therefore,

$$P_n^{\frac{k}{2},\sigma}\left(\frac{\sum(\phi_j^x)^2}{2}\right) = \frac{1}{2^n n!} : \left(\sum_1^k (\phi_j^x)^2\right)^n : . \tag{5.1}$$

Note that in particular, $: \sum_1^k (\phi_j^x)^2 :$ equals $(\phi_j^x)^2 - \sigma^x$ These variables are orthogonal in L^2. Let $\widetilde{l}^x = \hat{l}^x - \sigma^x$ be the centered occupation field. Note that an equivalent formulation of Theorem 2 is that the fields $\frac{1}{2} : \sum_1^k \phi_j^2 :$ and $\widetilde{\mathcal{L}}_{\frac{k}{2}}$ have the same law.

If we use complex fields $P_n^{\frac{k}{2},\sigma}\left(\frac{\sum \varphi_j^x \overline{\varphi}_j^x}{2}\right) = \frac{1}{2^n n!} : \left(\sum_1^k (\varphi_j^x \overline{\varphi}_j^x)\right)^n :$.

Let us now consider the relation of higher Wick powers with self intersection local times.

Recall that the renormalized n-th self intersections field $\widetilde{\mathcal{L}}_\alpha^{x,n} = P_n^{\alpha,\sigma}(\widehat{\mathcal{L}_\alpha}^x) = Q_n^{\alpha,\sigma}(\widetilde{\mathcal{L}_\alpha}^x)$ have been defined by orthonormalization in L^2 of the powers of the occupation time.

Then comes the

Proposition 19. (a) The fields $\widetilde{\mathcal{L}}_{\frac{k}{2}}^{\cdot,n}$ and $\frac{1}{n! 2^n} : \left(\sum_1^k \phi_j^2\right)^n :$ have the same law.

In particular $\widetilde{\mathcal{L}}_k^{\cdot,n}$ and $\frac{1}{n! 2^n} : \left(\sum_1^k \varphi_j \overline{\varphi}_j\right)^n :$ have the same law.

This follows directly from (5.1).

Remark 8. As a consequence, we obtain from (4.6) and (5.1) that:

$$\frac{k(k+2)...(k+2(n-1))}{2^n n!} = \sum_{n_1+...+n_k=n} \prod \frac{2n_i!}{(2^{n_i} n_i!)^2} \tag{5.2}$$

Moreover, it can be shown that:

$$\mathbb{E}\left(\prod_{j=1}^r Q_{k_j}^{\alpha,\sigma_{x_j}}(\widetilde{\mathcal{L}_\alpha}^{x_j})\right) = \sum_{\sigma \in \mathcal{S}_{k_1,k_2,...,k_j}} \alpha^{m(\sigma)} G^{y_1, y_{\sigma(1)}} ... G^{y_k, y_{\sigma(k)}}$$

$y_i = x_j$ for $\sum_1^{j-1} k_l + 1 \leq i \leq \sum_1^{j-1} k_l + k_j$ and where $\mathcal{S}_{k_1,k_2,...,k_j}$ denotes the set of permutations σ of $k = \sum k_j$ such that

$\sigma(\{\sum_1^{j-1} k_l + 1, ...\sum_1^{j-1} k_l + k_j\}) \cap \{\sum_1^{j-1} k_l + 1, ...\sum_1^{j-1} k_l + k_j\}$ is empty for all j.

The identity follows from Wick's theorem when α is an integer, then extends to all α since both members are polynomials in α. The condition on σ indicates that no pairing is allowed inside the same Wick power. For the proof, one can view each term of the form $: (\varphi_l^x \overline{\varphi}_l^x)^k :$ as the product of k distinct pairs, in a given order, then the pairings between φ's and $\overline{\varphi}$'s are defined by an element of $\mathcal{S}_{k_1,k_2,...k_j}$ and a system of permutations of $\mathcal{S}_{k_1}, ...\mathcal{S}_{k_j}$. This system of permutations produces multiplicities that cancel with the $\frac{1}{k_i!}$ factors in the expression. Note finally that $\mathbb{E}(\varphi_l^x \overline{\varphi}_l^y) = 2G^{x,y}$ to cancel the 2^{-k_i} factors.

5.3 The Gaussian Fock Space Structure

The Gaussian space $\mathcal{H}$ spanned by $\{\phi^x, x \in X\}$ is isomorphic to the dual of the Dirichlet space $\mathbb{H}^*$ by the linear map mapping ϕ^x on δ_x. This isomorphism extends into an isomorphism between the space of square integrable functionals of the Gaussian field and the real symmetric Fock space $\Gamma^{\odot}(\mathbb{H}^*) = \overline{\oplus \mathbb{H}^{*\odot n}}$ obtained as the closure of the sum of all symmetric tensor powers of $\mathbb{H}^*$ (the zero-th tensor power is $\mathbb{R}$). In the case of discrete spaces, these symmetric tensor powers can be represented simply as symmetric signed measures on X^n (with no additional constraint in the finite space case). In terms of the ordinary tensor product $\otimes$, the symmetric tensor product $\mu_1 \odot ... \odot \mu_n$ is defined as $\frac{1}{\sqrt{n!}} \sum_{\sigma \in \mathcal{S}_n} \mu_{\sigma(1)} \otimes ... \otimes \mu_{\sigma(n)}$ so that in particular, $\|\mu^{\odot n}\|^2 = n!(\sum G^{x,y}\mu_x\mu_y)^n$. The construction of $\Gamma(\mathbb{H}^*)$ is known as Bose second quantization. The isomorphism mentioned above is defined by the following identification, done for any μ in $\mathbb{H}^*$:

$$\exp^{\odot}(\mu) \rightleftarrows \exp(\sum_x \phi^x \mu_x - \frac{1}{2}\sum_{x,y} G^{x,y}\mu_x\mu_y)$$

which is an isometry as

$$\mathbb{E}_\phi(\exp(\sum_x \phi^x \mu_x - \frac{1}{2}\sum_{x,y} G^{x,y}\mu_x\mu_y)\exp(\sum_x \phi^x \mu'_x - \frac{1}{2}\sum_{x,y} G^{x,y}\mu'_x\mu'_y))$$

$$= \exp(-\sum_{x,y} G^{x,y}\mu_x\mu'_y).$$

The proof is completed by observing that linear combination of

$$\exp(\sum_x \phi^x \mu_x - \frac{1}{2}\sum_{x,y} G^{x,y}\mu_x\mu_y)$$

form a dense algebra in the space of square integrable functionals of the Gaussian field.

The n-th Wick power $: (\phi^x)^n :$ is the image of the n-th symmetric tensor power $\delta_x^{\odot n}$. More generally, for any μ in $\mathbb{H}^*$, $: (\sum_x \phi^x \mu_x)^n :$ is the image of the n-th symmetric tensor power $\mu^{\odot n}$. Therefore, $: \phi^{x_1}\phi^{x_2}...\phi^{x_n} :$ is the image of $\delta_{x_1} \odot ... \odot \delta_{x_n}$, and polynomials of the field are associated with the non completed Fock space $\oplus \mathbb{H}^{*\odot n}$.

For any $x \in X$, the annihilation operator a_x and the creation operator a_x^* are defined as follows, on the uncompleted Fock space $\oplus \mathbb{H}^{* \odot n}$:

$$a_x(\mu_1 \odot \ldots \odot \mu_n) = \sum_k G\mu_k(x)\mu_1 \odot \ldots \odot \mu_{k-1} \odot \mu_{k+1} \odot \ldots \odot \mu_n$$

$$a_x^*(\mu_1 \odot \ldots \odot \mu_n) = \delta_x \odot \mu_1 \odot \ldots \odot \mu_n.$$

Moreover, we set $a_x 1 = 0$ for all x. These operator a_x and a_x^* are clearly dual of each other and verify the commutation relations:

$$[a_x, a_y^*] = G^{x,y} \quad [a_x^*, a_y^*] = [a_x, a_y] = 0.$$

The isomorphism allows to represent them on polynomials of the field as follows:

$$a_x \rightleftarrows \sum_y G^{x,y} \frac{\partial}{\partial \phi^y}$$

$$a_x^* \rightleftarrows \phi_x - \sum_y G^{x,y} \frac{\partial}{\partial \phi^y}$$

Therefore, the Fock space structure is entirely transported on the space of square integrable functionals of the free field.

In the case of a complex field φ, the space of square integrable functionals of φ and $\overline{\varphi}$ is isomorphic to the tensor product of two copies of the symmetric Fock space $\Gamma^{\odot}(\mathbb{H}^*)$, we will denote by $\mathcal{F}_B$. It is isomorphic to $\Gamma^{\odot}(\mathbb{H}_1^* + \mathbb{H}_2^*)$, $\mathbb{H}_1$ and $\mathbb{H}_2$ being two copies of $\mathbb{H}$. The complex Fock space structure is defined by two commuting sets of creation and annihilation operators:

$$a_x = \sqrt{2} \sum_y G^{x,y} \frac{\partial}{\partial \varphi^y} \quad a_x^* = \frac{\varphi^x}{\sqrt{2}} - \sqrt{2} \sum_y G^{x,y} \frac{\partial}{\partial \overline{\varphi}^y}$$

$$b_x = \sqrt{2} \sum_y G^{x,y} \frac{\partial}{\partial \overline{\varphi}^y} \quad b_x^* = \frac{\overline{\varphi}^x}{\sqrt{2}} - \sqrt{2} \sum_y G^{x,y} \frac{\partial}{\partial \varphi^y}$$

(Recall that if $z = x + iy$, $\frac{\partial}{\partial z} = \frac{\partial}{\partial x} - i\frac{\partial}{\partial y}$ and $\frac{\partial}{\partial \overline{z}} = \frac{\partial}{\partial x} + i\frac{\partial}{\partial y}$).

We have

$$\varphi^x = \sqrt{2}(b_x + a_x^*) \text{ and } \overline{\varphi}^x = \sqrt{2}(a_x + b_x^*)$$

See [35, 48] for a more general description of this isomorphism.

Exercise 22. Let A and B be two polynomials in φ and $\overline{\varphi}$, identified with finite degree element in $\mathcal{F}_B$. Show by recurrence on the degrees that $AB = \sum \frac{1}{p!} \gamma_p(A, B)$ with $\gamma_0(A, B) = A \odot B$ and $\gamma_{p+1}(A, B) = \sum_x (\gamma_p(a_x A, b_x B) + \gamma_p(b_x A, a_x B))$.

5.4 The Poissonian Fock Space Structure

Another symmetric Fock space structure is defined on the spaces of L^2-functionals of the loop ensemble $\mathcal{LP}$. It is based on the space $\mathfrak{h} = L^2(\mu^L)$ where μ^L denotes $\mu \otimes Leb$. For any $G \in L^2(\mu^L)$ define $G_{(\varepsilon)}(t,l) = G(t,l)1_{\{T(l)>\varepsilon\}}1_{\{|t|<\frac{1}{\varepsilon}\}}$. Note that $G_{(\varepsilon)}$ is always integrable. Define

$$\mathfrak{h}_0 = \bigcup_{\varepsilon>0}\{G \in L^\infty(\mu^L), \exists \varepsilon > 0,\ G = G_{(\varepsilon)} \text{ and } \int G d(\mu^L) = 0\}.$$

The algebra $\mathfrak{h}_0$ is dense in $L^2(\mu^L)$ (as, for example, compactly supported square integrable functions with zero integral are dense in $L^2(Leb)$).

Given any F in $\mathfrak{h}_0$, $\mathcal{LP}(F) = \sum_{(t_i,l_i)\in\mathcal{LP}} F(t_i,l_i)$ is well defined and $\mathbb{E}(\mathcal{LP}(F)^2) = \langle F,F\rangle_{L^2(\mu^L)}$. By Stone Weierstrass theorem, the algebra generated by $\mathcal{LP}(\mathfrak{h}_0)$ is dense in $L^2(\mu^L)$.

By Campbell formula, the n-th chaos, isomorphic to the symmetric tensor product $\mathfrak{h}^{\odot n}$, can be defined as the closure of the linear span of functions of n distinct points of $\mathcal{LP}$ of the form

$$\sum_{\sigma\in\mathcal{S}_n} \prod_1^n G_{\sigma(j)}(l_j,\alpha_j)$$

with G_j in $\mathfrak{h}_0$.

Denote by $t\mathcal{LP}$ the jump times of the Poisson process $\mathcal{LP}$. It follows directly from formula (4.2) that for

$$\Phi = \frac{1}{\sqrt{n!}}\sum_{\sigma\in\mathcal{S}_n}\sum_{t_1<t_2<...<t_n\in t\mathcal{LP}}\prod_1^n G_{\sigma(j)}(l_j,t_j)) \text{ and } \Phi'$$

$$= \frac{1}{\sqrt{n!}}\sum_{\sigma\in\mathcal{S}_{n'}}\sum_{t_1<t_2...<t_{n'}\in t\mathcal{LP}}\prod_1^{n'} G'_{\sigma(j')}(l_{j'},t_{j'})),$$

with G_j, $G'_{j'}$ in $\mathfrak{h}_0$,

$$\mathbb{E}(\Phi\Phi') = 1_{\{n=n'\}}Per(\langle G_j,G'_{j'}\rangle_{L^2(\mu^L)},1 \le j,j' \le n)$$

which equals $1_{\{n=n'\}}\langle G_1\odot G_2...\odot G_n,G'_1\odot G'_2...\odot G'_{n'}\rangle$, $\odot$ denoting the symmetric tensor product (Cf. [3,35]).

This proves the existence of an isomorphism Iso between the algebra generated by $\mathcal{LP}(\mathfrak{h}_0)$ and the tensor algebra $\oplus\mathfrak{h}_0^{\odot n}$ which extends into an isomorphism between the space $L^2(\mathbb{P}_{\mathcal{LP}})$ of square integrable functionals of the Poisson process of loops $\mathcal{LP}$ and the symmetric (or bosonic) Fock space $\oplus \mathfrak{h}^{\odot n}$. We have

$$Iso(\frac{1}{\sqrt{n!}} \sum_{\sigma \in \mathcal{S}_n} \sum_{t_1 < t_2 < ... < t_n \in t\mathcal{LP}} \prod_1^n G_{\sigma(j)}(l_j, t_j))) = G_1 \odot G_2 \odot ... \odot G_n.$$

This formula extends to $G_i \in \mathfrak{h}$. The closure of the space that functionals of this form generate linearly is by definition the n-th chaos which is isomorphic to the symmetric tensor product $\mathfrak{h}^{\odot n}$.

Note that for any G in $\mathfrak{h}_0$, the image by this isomorphism Iso of the tensor exponential $\exp^{\odot}(G)$ is $\prod_{t_i \in t\mathcal{LP}}(1 + G(l_i, t_i))$.

Note also that for all F in $\mathfrak{h}$, $\|\exp^{\odot}(F)\|^2 = e^{\int F^2 d\mu^L}$

Proposition 20. *For any F in $\mathfrak{h}$, the image by Iso of the tensor exponential $\exp^{\odot}(F)$ is obtained as the limit in L^2, as $\varepsilon \to 0$ of*

$$\prod_{t_i \in t\mathcal{LP}}(1 + F_{(\varepsilon)}(l_i, t_i))e^{-\int F_{(\varepsilon)} d\mu^L}.$$

Proof. Note first that

$$\mathbb{E}(\prod_{t_i \in t\mathcal{LP}}(1 + F_{(\varepsilon)}(l_i, t_i))e^{-\int F_{(\varepsilon)} d\mu^L}) = 1$$

and

$$\mathbb{E}([\prod_{t_i \in t\mathcal{LP}}(1 + F_{(\varepsilon)}(l_i, t_i))e^{-\int F_{(\varepsilon)} d\mu^L}]^2)$$

$$= \exp(\int [(1 + F_{(\varepsilon)})^2 - 1]d\mu^L)e^{-2\int F_{(\varepsilon)} d\mu^L}$$

$$= \exp(\int F_{(\varepsilon)}^2 d\mu^L)$$

converges towards $\exp(\int F^2 d\mu^L)$.

Then note that for any G in $\mathfrak{h}_0$,

$$\lim_{\varepsilon \to 0} \mathbb{E}(\prod_{t_i \in t\mathcal{LP}}(1 + G(l_i, t_i))(1 + F_{(\varepsilon)}(l_i, t_i))e^{-\int F_{(\varepsilon)} d\mu^L})$$

$$= \lim_{\varepsilon \to 0} \exp(\int [[1 + G][1 + F_{(\varepsilon)}] - 1]e^{-\int (F_{(\varepsilon)})d\mu^L} d(\mu^L))$$

$$= \lim_{\varepsilon \to 0} \exp(\int F_{(\varepsilon)} G d\mu^L) = \exp(\int F G d\mu^L) = \langle \exp^{\odot}(F), \exp^{\odot}(G) \rangle.$$

$\square$

Exercise 23. For any F in $\mathfrak{h}$, set $F_{\leq \alpha}(l,t) = F(l,t)1_{\{t\leq\alpha\}}$. Show that $Iso(\exp^{\odot}(F_{\leq\alpha}))$ is a $\sigma(\mathcal{L}_\alpha)$-martingale.

Prove that the $\sigma(\mathcal{L}_\alpha)$-martingale $(1 + \sigma_x t)^{-\alpha}\exp(\frac{\widehat{\mathcal{L}_\alpha}^x t}{1+\sigma_x t})$ is in this way associated with $F(l,t) = e^{\frac{\hat{l}^x t}{1+\sigma_x t}} - 1$.

Deduce from this an expression of $P_k^{\alpha,\sigma_x}(\widehat{\mathcal{L}_\alpha}^x)$ in terms of $\sigma_x^{k_i}D_{k_i}(\frac{\hat{l}_i^x}{\sigma_x})$ (the polynomials defined in Sect. 2.4), l_i denoting distinct loops in $\mathcal{L}_\alpha$ and k_i positive integers less than k.

For any G in $\mathfrak{h}_0$, unbounded annihilation and creation operators $\mathfrak{A}_G$ and $\mathfrak{A}_G^*$ are defined on $\oplus\mathfrak{h}^{\odot n}$

$$\mathfrak{A}_G(G_1 \odot ... \odot G_n) = \sum_{k=1}^{n} \langle G, G_k\rangle_{\mathfrak{h}} \, G_1 \odot ... G_{k-1} \odot G_{k+1} ... \odot G_n$$

and

$$\mathfrak{A}_G^*(G_1 \odot ... \odot G_n) = G \odot G_1 \odot ... \odot G_n$$

Note that

$$[\mathfrak{A}_G^*, \mathfrak{A}_F^*] = [\mathfrak{A}_G, \mathfrak{A}_F] = 0$$

$$[\mathfrak{A}_G, \mathfrak{A}_F^*] = \langle F, G\rangle_{L^2(\mu^L)}$$

Moreover, $\mathfrak{A}_G^*$ is adjoint to $\mathfrak{A}_G$ in $L^2(\mathbb{P}_{\mathcal{LP}})$, and the operators $\mathfrak{F}_G = \mathfrak{A}_G^* + \mathfrak{A}_G$ commute.

Note also that the creation operator can be defined directly on the space of loop configurations: by Proposition 15 given any bounded functional Ψ on loops configurations,

$$Iso\mathfrak{A}_G^* Iso^{-1}\Psi(\mathcal{LP}) = \int \Psi(\mathcal{LP} \cup \{l,t\})G(l,t)d\mu^L$$

It is enough to verify it for $Iso^{-1}\Psi$ in $\mathfrak{h}_0^{\odot n}$.

For any G in $\mathfrak{h}_0 \cap L^\infty$, note that $\mathfrak{F}_G$ does not represent the multiplication by $\mathcal{LP}(G)$, though we have, for all Φ in $\oplus\mathfrak{h}^{\odot n}$ and G in $\mathfrak{h}_0$, $\mathbb{E}(Iso(\Phi)\mathcal{LP}(G)) = \langle \mathfrak{F}_G 1, \Phi\rangle = \langle 1, \mathfrak{F}_G\Phi\rangle$.

The representation of this operator of multiplication in the Fock space structure can be done as follows:

Setting $M_G F = GF$, for all Φ in $\oplus\mathfrak{h}^{\odot n}$

$$\left(\sum_{t_i\in t\mathcal{LP}} G(l_i,t_i)\right)Iso(\Phi) = Iso(\mathfrak{F}_G\Phi + d\Gamma(M_G)).$$

The notation $d\Gamma$ refers to the second quantisation functor Γ: if B is any bounded operator on a Hilbert space $\mathfrak{h}$, $\Gamma(B)$ is defined on $\oplus\mathfrak{h}^{\odot n}$ by the sum of the operators $B^{\otimes n}$ acting on each symmetric tensor product $\mathfrak{h}^{\odot n}$ and $d\Gamma(B) = \frac{\partial}{\partial t}\Gamma(e^{tB})|_{t=0}$. In fact, given any orthonormal basis E_k of $\mathfrak{h}$, we have, for any Φ in $\oplus \mathfrak{h}^{\odot n}$ and G in $\mathfrak{h}$

$$d\Gamma(B) = \sum \mathfrak{A}^*_{BE_k}\mathfrak{A}_{E_k} = \sum \mathfrak{A}^*_{E_k}\mathfrak{A}_{B^*E_k}$$

as

$$\sum \langle F, E_k\rangle_{L^2(\mu^L)}(BE_k) = \sum \langle F, BE_k\rangle_{L^2(\mu^L)} E_k = BF.$$

Chapter 6
Energy Variation and Representations

6.1 Variation of the Energy Form

The loop measure μ depends on the energy e which is defined by the free parameters C, κ. It will sometimes be denoted μ_e. We shall denote $\mathcal{Z}_e$ the determinant $\det(G) = \det(M_\lambda - C)^{-1}$. Then $\mu(p > 1) = \log(\mathcal{Z}_e) + \sum_{x \in X} \log(\lambda_x)$.

$\mathcal{Z}_e^\alpha$ is called the partition function of $\mathcal{L}_\alpha$.

We wish to study the dependance of μ on C and κ. The following result is suggested by an analogy with quantum field theory (Cf. [11]).

Proposition 21. (i) $\frac{\partial \mu}{\partial \kappa_x} = -\widehat{l^x} \mu$.

(ii) If $C_{x,y} > 0$, $\frac{\partial \mu}{\partial C_{x,y}} = -T^{x,y}\mu$ with $T^{x,y}(l) = (\widehat{l^x} + \widehat{l^y}) - \frac{N_{x,y}}{C_{x,y}}(l) - \frac{N_{y,x}}{C_{x,y}}(l)$.

Proof. Recall that by formula (2.8): $\mu^*(p = 1, \xi = x, \widehat{\tau} \in dt) = e^{-\lambda_x t}\frac{dt}{t}$ and

$$\mu^*(p = k, \xi_i = x_i, \widehat{\tau}_i \in dt_i) = \frac{1}{k}\prod_{x,y}C_{x,y}^{N_{x,y}}\prod_x \lambda_x^{-N_x}\prod_{i \in \mathbb{Z}/p\mathbb{Z}}\lambda_{\xi_i}e^{-\lambda_{\xi_i}t_i}dt_i.$$

Moreover we have $C_{x,y} = C_{y,x} = \lambda_x P_y^x$ and $\lambda_x = \kappa_x + \sum_y C_{x,y}$.
The two formulas follow by elementary calculation. $\qquad \square$

Recall that $\mu(\widehat{l^x}) = G^{x,x}$ and $\mu(N_{x,y}) = G^{x,y}C_{x,y}$. So we have $\mu(T^{x,y}) = G^{x,x} + G^{y,y} - 2G^{x,y}$. Then, the above proposition allows us to compute all moments of T and $\widehat{l}$ relative to μ_e (they could be called Schwinger functions).

Exercise 24. Use the proposition above to show that:

$$\int \widehat{l^x}\widehat{l^y}\mu(dl) = (G^{x,y})^2$$

$$\int \widehat{l^x}T^{y,z}(l)\mu(dl) = (G^{x,y} - G^{x,z})^2$$

Y. Le Jan, *Markov Paths, Loops and Fields*, Lecture Notes in Mathematics 2026, DOI 10.1007/978-3-642-21216-1_6, © Springer-Verlag Berlin Heidelberg 2011

and

$$\int T^{x,y}(l) T^{u,v}(l) \mu(dl) = (G^{x,u} + G^{y,v} - G^{x,v} - G^{y,u})^2 = (K^{(x,y),(u,v)})^2$$

Hint: The calculations are done noticing that for any invertible matrix function $M(s)$, $\frac{d}{ds} M(s)^{-1} = -M(s)^{-1} M'(s) M(s)^{-1}$. The formula is applied to $M = M_\lambda - C$ and $s = \kappa_x$ or $C_{x,y}$.

Exercise 25. Show that $\int (\frac{1}{2} \sum_{x,y} C_{x,y} T^{x,y}(l) + \sum_x \kappa_x \widehat{l^x}) \mu(dl) = |X|$ and that more generally, for any $D \subset X$,

$$\int (\frac{1}{2} \sum_{x,y} C_{x,y \in D} T^{x,y}(l) + \sum_{x \in D} \kappa_x \widehat{l^x}) \mu(dl) = |D| + \sum_{x \in D, y \in X-D} C_{x,y} G^{x,y}$$

Set

$$T_{x,y}^{(\alpha)} = \sum_{l \in \mathcal{L}_\alpha} T_{x,y}(l) = (\widehat{\mathcal{L}_\alpha^x} + \widehat{\mathcal{L}_\alpha^y}) - \frac{N_{x,y}^{(\alpha)}}{C_{x,y}} - \frac{N_{y,x}^{(\alpha)}}{C_{x,y}}$$

and

$$\tilde{T}_{x,y}^{(\alpha)} = T_{x,y}^{(\alpha)} - \mathbb{E}(T_{x,y}^{(\alpha)}) = T_{x,y}^{(\alpha)} - \alpha(G^{x,x} + G^{y,y} - 2G^{x,y}).$$

We can apply Proposition 21 to the Poissonnian loop ensembles, to get the following

Corollary 4. *For any bounded functional Φ on loop configurations*

(i) $\frac{\partial}{\partial \kappa_x} \mathbb{E}(\Phi(\mathcal{L}_\alpha)) = -\mathbb{E}(\Phi(\mathcal{L}_\alpha) \widehat{\mathcal{L}_\alpha^x}) = \alpha \int \mathbb{E}((\Phi(\mathcal{L}_\alpha) - \Phi(\mathcal{L}_\alpha \cup \{\gamma\})) \mu^{x,x}(d\gamma).$
(ii) If $C_{x,y} > 0$,

$$\frac{\partial}{\partial C_{x,y}} \mathbb{E}(\Phi(\mathcal{L}_\alpha)) = -\mathbb{E}(\tilde{T}_{x,y}^{(\alpha)} \Phi(\mathcal{L}_\alpha))$$

$$= \alpha \int \mathbb{E}((\Phi(\mathcal{L}_\alpha) - \Phi(\mathcal{L}_\alpha \cup \{\gamma\}))[\mu^{x,x}(d\gamma) + \mu^{y,y}(d\gamma) - \mu^{x,y}(d\gamma) - \mu^{y,x}(d\gamma)].$$

The proof is easily performed, taking first Φ of the form $\sum_{l_1 \neq l_2 \cdots \neq l_q \in \mathcal{L}_\alpha} \prod_1^q G_j(l_j)$. We apply Campbell formula to deduce the first half of both identities, then Corollary 2 to get the second half.

This result should be put in relation with Propositions 4 and 10 and with the Poissonian Fock space structure defined above.

Exercise 26. Show that using Theorem 2, Corollary 4 implies that for any function F of an non-negative field and any edge (x_i, y_i):

$$\mathbb{E}_\phi(\frac{1}{2} : (\phi^x - \phi^y)^2 : F(\frac{1}{2}\phi^2)) = \int \mathbb{E}(F(\widehat{\mathcal{L}_{\frac{1}{2}}}) \tilde{T}_{x,y}^{(\frac{1}{2})})$$

$$\mathbb{E}_\phi(: \phi^x \phi^y : F(\frac{1}{2}\phi^2)) = -\int \mathbb{E}(F(\widehat{\mathcal{L}_{\frac{1}{2}}}) N_{x,y})$$

Hint: Express the Gaussian measure and use the fact that $-\frac{\partial}{\partial C_{x,y}}$ $Log(\det(G)) = G^{x,x} + G^{y,y} - 2G^{x,y}$

Exercise 27. Setting $\widetilde{l}^x = \widehat{l}^x - G^{x,x}$ and $\widetilde{T}_{x,y}(l) = T_{x,y}(l) - (G^{x,x} + G^{y,y} - 2G^{x,y})$, show that we have:

$$\int \widetilde{l}_x \widetilde{l}_y \mu(dl) = (G^{x,y})^2 - G^{x,x}G^{y,y} = \det \begin{pmatrix} \langle \phi^x, \phi^x \rangle & \langle \phi^x, \phi^y \rangle \\ \langle \phi^y, \phi^x \rangle & \langle \phi^y, \phi^y \rangle \end{pmatrix}$$

$$\int \widetilde{l}_x \widetilde{T}_{y,z}(l)\mu(dl) = (G^{x,y} - G^{x,z})^2 - G^{x,x}(G^{z,z} + G^{y,y} - 2G^{y,z})$$

$$= \det \begin{pmatrix} \langle \phi^x, \phi^x \rangle & \langle \phi^x, \phi^y - \phi^z \rangle \\ \langle \phi^x, \phi^y - \phi^z \rangle & \langle \phi^y, \phi^y \rangle \end{pmatrix}$$

$$\int \widetilde{T}_{x,y}(l)\widetilde{T}_{u,v}(l)\mu(dl) = (G_{x,u} + G_{y,v} - G_{x,v} - G_{y,u})^2$$

$$- (G^{x,x} + G^{y,y} - 2G^{x,y})(G^{u,u} + G^{v,v} - 2G^{u,v})$$

$$= -\det \begin{pmatrix} \langle \phi^x - \phi^y, \phi^x - \phi^y \rangle & \langle \phi^x - \phi^y, \phi^u - \phi^v \rangle \\ \langle \phi^u - \phi^v, \phi^x - \phi^y \rangle & \langle \phi^u - \phi^v, \phi^u - \phi^v \rangle \end{pmatrix}.$$

Exercise 28. For any bounded functional Φ on loop configurations, give two different expressions for $\frac{\partial^2}{\partial \kappa_x \partial \kappa_y}\mathbb{E}(\Phi(\mathcal{L}_\alpha))$, $\frac{\partial^2}{\partial C_{x,y}\partial \kappa_z}\mathbb{E}(\Phi(\mathcal{L}_\alpha))$ and $\frac{\partial^2}{\partial C_{x,y}\partial C_{u,v}}\mathbb{E}(\Phi(\mathcal{L}_\alpha))$.

The Proposition 21 is in fact the infinitesimal form of the following formula.

Proposition 22. *Consider another energy form e' defined on the same graph. Then we have the following identity:*

$$\frac{\partial \mu_{e'}}{\partial \mu_e} = e^{\sum N_{x,y}\log(\frac{C'_{x,y}}{C_{x,y}}) - \sum(\lambda'_x - \lambda_x)\widehat{l}^x}.$$

Consequently

$$\mu_e((e^{\sum N_{x,y}\log(\frac{C'_{x,y}}{C_{x,y}}) - \sum(\lambda'_x - \lambda_x)\widehat{l}^x} - 1)) = \log(\frac{\mathcal{Z}_{e'}}{\mathcal{Z}_e}). \quad (6.1)$$

Proof. The first formula is a straightforward consequence of (2.6). The proof of (6.1) goes by evaluating separately the contribution of trivial loops, which equals $\sum_x \log(\frac{\lambda_x}{\lambda'_x})$. Indeed,

$$\mu_e((e^{\sum N_{x,y}\log(\frac{C'_{x,y}}{C_{x,y}}) - \sum(\lambda'_x - \lambda_x)\widehat{l}^x} - 1) = \mu_{e'}(p > 1) - \mu_e(p > 1)$$

$$+ \mu_e(1_{\{p=1\}}(e^{\sum(\lambda'_x - \lambda_x)\widehat{l}^x} - 1)).$$

The difference of the first two terms equals $\log(\mathcal{Z}_{e'}) + \sum \log(\lambda'_x) -$ $(\log(\mathcal{Z}_e) - \sum \log(\lambda_x))$. The last term equals $\sum_x \int_0^\infty (e^{-\frac{\lambda'_x - \lambda_x}{\lambda_x}t} - 1)\frac{e^{-t}}{t} dt$ which can be computed as before:

$$\mu_e(1_{\{p=1\}}(e^{\sum (\lambda'_x - \lambda_x)\widehat{l}^x} - 1)) = -\sum \log(\frac{\lambda'_x}{\lambda_x}) \tag{6.2}$$

$\square$

Integrating out the holding times, formula (6.1) can be written equivalently:

$$\mu_e(\prod_{(x,y)} [\frac{C'_{x,y}}{C_{x,y}}]^{N_{x,y}} \prod_x [\frac{\lambda_x}{\lambda'_x}]^{N_x+1} - 1) = \log(\frac{\mathcal{Z}_{e'}}{\mathcal{Z}_e}) \tag{6.3}$$

and therefore

$$\mathbb{E}(\prod_{(x,y)} [\frac{C'_{x,y}}{C_{x,y}}]^{N_{x,y}(\mathcal{L}_\alpha)} \prod_x [\frac{\lambda_x}{\lambda'_x}]^{N_x(\mathcal{L}_\alpha)+1}) = (\frac{\mathcal{Z}_{e'}}{\mathcal{Z}_e})^\alpha \tag{6.4}$$

or equivalently

$$\mathbb{E}(\prod_{(x,y)} [\frac{C'_{x,y}}{C_{x,y}}]^{N_{x,y}(\mathcal{L}_\alpha)} e^{-\langle \lambda'-\lambda, \widehat{\mathcal{L}_\alpha}\rangle}) = (\frac{\mathcal{Z}_{e'}}{\mathcal{Z}_e})^\alpha \tag{6.5}$$

Note also that $\prod_{(x,y)} [\frac{C'_{x,y}}{C_{x,y}}]^{N_{x,y}} = \prod_{\{x,y\}} [\frac{C'_{x,y}}{C_{x,y}}]^{N_{x,y}+N_{y,x}}$.

Remark 9. These $\frac{\mathcal{Z}_{e'}}{\mathcal{Z}_e}$ determine, when e' varies with $\frac{C'}{C} \le 1$ and $\frac{\lambda'}{\lambda} = 1$, the Laplace transform of the distribution of the traversal numbers of non oriented links $N_{x,y} + N_{y,x}$.

Remark 10. (h-transforms) Note that if $C'_{x,y} = h^x h^y C_{x,y}$ and $\kappa'_x = -h^x (Lh)^x \lambda_x$ for some positive function h on E such that $Lh \le 0$, as $\lambda' = h^2\lambda$ and $[P']^x_y = \frac{1}{h^x} P^x_y h^y$, we have $[G']^{x,y} = \frac{G^{x,y}}{h^x h^y}$ and $\frac{\mathcal{Z}_{e'}}{\mathcal{Z}_e} = \frac{1}{\prod (h^x)^2}$.

Remark 11. Note also that $[\frac{\mathcal{Z}_{e'}}{\mathcal{Z}_e}]^{\frac{1}{2}} = \mathbb{E}_\phi(e^{-\frac{1}{2}[e'-e](\phi)})$, if ϕ is the Gaussian free field associated with e.

6.2   One-Forms and Representations

Other variables of interest on the loop space are associated with elements of the space $\mathbb{A}^-$ of odd real valued functions ω on oriented links: $\omega^{x,y} = -\omega^{y,x}$. Let us mention a few elementary results.

The operator $[P^{(\omega)}]^x_y = P^x_y \exp(i\omega^{x,y})$ is also self adjoint in $L^2(\lambda)$. The associated loop variable can be written $\sum_{x,y} \omega^{x,y} N_{x,y}(l)$. We will denote it $\int_l \omega$. This notation will be used even when ω is not odd. Note that $\int_l \omega$ is invariant if ω is replaced by $\omega + dg$ for some g. Set $[G^{(\omega)}]^{x,y} = \frac{[(I-P^{(\omega)})^{-1}]^x_y}{\lambda_y}$. By an argument similar to the one given above for the occupation field, we have:

$$\mathbb{P}^t_{x,x}(e^{i\int_l \omega} - 1) = \exp(t(P^{(\omega)} - I))^x_x - \exp(t(P - I))^x_x.$$

Integrating in t after expanding, we get from the definition of μ:

$$\int (e^{i\int_l \omega} - 1)d\mu(l) = \sum_{k=1}^{\infty} \frac{1}{k}[Tr((P^{(\omega)})^k) - Tr((P)^k)].$$

Hence $\int (e^{i\int_l \omega} - 1)d\mu(l) = \log[\det(-L(I - P^{(\omega)})^{-1})] = \log(\det(G^{(\omega)}G^{-1}))$

We can now extend the previous formulas (6.3) and (6.4) to obtain, setting $\det(G^{(\omega)}) = \mathcal{Z}_{e,\omega}$

$$\int (e^{\sum N_{x,y} \log(\frac{C'_{x,y}}{C_{x,y}}) - \sum(\lambda'_x - \lambda_x)\hat{l}_x + i\int_l \omega} - 1)\mu_e(dl) = \log(\frac{\mathcal{Z}_{e',\omega}}{\mathcal{Z}_e}) \qquad (6.6)$$

and

$$\mathbb{E}(\prod_{x,y}[\frac{C'_{x,y}}{C_{x,y}}e^{i\omega_{x,y}}]^{N^{(\alpha)}_{x,y}} e^{-\sum(\lambda'_x - \lambda_x)\hat{\mathcal{L}}_\alpha^x}) = (\frac{\mathcal{Z}_{e',\omega}}{\mathcal{Z}_e})^\alpha \qquad (6.7)$$

Remark 12. The α-th power of a complex number is a priori not univoquely defined as a complex number. But $\log[\det(I - P^{(\omega)})]$ and therefore $\log(\mathcal{Z}_{e,\omega})$ are well defined as $P^{(\omega)}$ is a contraction. Then $\mathcal{Z}^\alpha_{e,\omega}$ is taken to be $\exp(\alpha \log(\mathcal{Z}_{e,\omega}))$.

Remark 13. Note also that if $\varphi = \phi_1 + i\phi_2$ is the complex Gaussian free field associated with e, and if we set $e'^{,(\omega)}(\varphi) = \frac{1}{2}[\sum(\lambda'_x - \lambda_x)\varphi^x\overline{\varphi}^x - \sum C'_{x,y}e^{i\omega_{x,y}}\varphi^x\overline{\varphi}^y]$,

$$\frac{\mathcal{Z}_{e',\omega}}{\mathcal{Z}_e} = \mathbb{E}_\varphi(e^{-\frac{1}{2}[e'^{,(\omega)} - e](\varphi)})$$

To simplify the notations slightly, one could consider more general energy forms with complex valued conductances so that the discrete one form is included in e'. But it is more interesting to generalize the notion of perturbation of P into $P^{(\omega)}$ as follows:

Definition 1. A unitary representation of the graph (X, E) is a family of unitary matrices $[U^{x,y}]$, with common rank d_U, indexed by E^O, such that $[U^{y,x}] = [U^{x,y}]^{-1}$.

We set $P^{(U)} = P \otimes U$ (more explicitly $[P^{(U)}]^{y,j}_{x,i} = P^y_x [U^{x,y}]^j_i$).

Similarly, we can define $C^{(U)} = \frac{\lambda}{d_U} P^{(U)}$, $V^{(U)} = (I - P^{(U)})^{-1}$, $G^{(U)} = \frac{d_U V^{(U)}}{\lambda}$. One should think of these matrices as square matrices indexed by X, whose entries are multiples of elements of $SU(d_U)$.

One forms define one-dimensional representations. The sum and tensor product of two unitary representations U and V are unitary representations are defined as usual, and their ranks are respectively $d_U + d_V$ and $d_U d_V$.

Definition 2. Given any based loop l, if $p(l) \geq 2$ and the associated discrete based loop is $\xi = (\xi_1, \xi_2, \ldots, \xi_p)$, set $\tau_U(l) = \frac{1}{d_U} Tr(\prod U^{\xi_i, \xi_{i+1}})$, and $\tau_U(l) = 1$ if $p(l) = 1$.

For any set of loops $\mathcal{L}$, we set $\tau_U(\mathcal{L}) = \prod_{l \in \mathcal{L}} \tau_U(l)$.

Remark 14. (a) $|\tau_U(l)| \leq 1$.

(b) τ_U is obviously a functional of the discrete *loop* $\overset{\circ}{\xi}$ contained in $\overset{\circ}{l}$.

(c) $\tau_U(l) = 1$ if $\overset{\circ}{\xi}$ is tree-like. In particular it is always the case when the graph is a tree.

(d) If U and V are two unitary representations of the graph, $\tau_{U+V} = \tau_U + \tau_V$ and $\tau_{U \otimes V} = \tau_U \tau_V$.

From (b) and (c) above it is easy to get the first part of.

Theorem 3. *(i) The trace $\tau_U(l)$ depends only on the canonical geodesic loop associated with the loop $\overset{\circ}{\xi}$, i.e. of the conjugacy class of the element of the fundamental group defined by the based loop ξ.*

(ii) The variables $\tau_U(l)$ determine, as U varies, the geodesic loop associated with l.

Proof. The second assertion follows from the fact that traces of unitary representations separate the conjugacy classes of finite groups (Cf. [42]) and from the so-called CS-property satisfied by free groups (Cf. [51]): given two elements belonging to different conjugacy classes, there exists a finite quotient of the group in which they are not conjugate.

Let us fix a base point x_0 in X and a spanning tree T. An oriented edge (x, y) which is not in T defines an element $\gamma_{x,y}$ of the fundamental group Γ_{x_0}, with $\gamma_{y,x} = \gamma_{x,y}^{-1}$. For eny edge $(u, v) \in T$, we set $\gamma_{u,v} = I$. For any discrete based loop $l = (x_1, x_2, \ldots, x_p)$, set $\gamma_l = \gamma_{x_1,x_2} \cdots \gamma_{x_{p-1},x_p} \gamma_{x_p,x_1}$. Then, if two based loops l_1 and l_2 define distinct geodesic loops, there exists a finite quotient $G = \Gamma_{x_0}/H$ of Γ_{x_0} in which the classes of their representatives γ_{l_i} are not conjugate. Denote by $\overline{\gamma}$ the class of γ in G. Then there exists a unitary representation ρ of G such that $Tr(\rho(\gamma_{l_1})) \neq Tr(\rho(\gamma_{l_2}))$. Then take $U_{x,y} = \rho(\overline{\gamma}_{x,y})$. We see that $\tau_U(l_1) \neq \tau_U(l_2)$. $\square$

Again, by an argument similar to the one given for the occupation field, we have:

$$\mathbb{P}^t_{x,x}(\tau_U - 1) = \frac{1}{d_U} \sum_{i=1}^{d_U} \exp(t(P^{(U)} - I))^{x,i}_{x,i} - \exp(t(P - I))^x_x.$$

Integrating in t after expanding, we get from the definition of μ:

$$\int (\tau_U(l) - 1)d\mu(l) = \sum_{k=1}^{\infty} \frac{1}{k}[\frac{1}{d_U}Tr((P^{(U)})^k) - Tr((P)^k)].$$

We can extend P into a matrix $P^{(I_{d_U})} = P \otimes I_{d_U}$ indexed by $X \times \{1,\ldots,d_U\}$ by taking its tensor product with the identity on $\mathbb{R}^{d_U}$.
Then:

$$\int (\tau_U(l) - 1)d\mu(l) = \frac{1}{d_U} \sum_{k=1}^{\infty} \frac{1}{k}[Tr((P^{(U)})^k) - Tr((P^{(I_{d_U})})^k)].$$

Hence, as in the case of the occupation field

$$\int (\tau_U(l) - 1)d\mu(l) = \frac{1}{d_U} \log(\det(V^{(U)})[V \otimes I_{d_U}]^{-1})) = \frac{1}{d_U} \log(\det(G^{(U)})) - \log(\det(G))$$

as $\det(G \otimes I_{d_U}) = \det(G)^{d_U}$.

Then, denoting $\mathcal{Z}_{e,U}$ the $\left(\frac{1}{d_U}\right)$-th power of the determinant of the $(|X|\,d_U, |X|\,d_U)$ matrix $G^{(U)}$ (well defined by Remark 12), the formulas (6.6) and (6.7) extend easily to give the following

Proposition 23. (a) $\int (e^{\sum N_{x,y} \log(\frac{C'_{x,y}}{C_{x,y}}) - \sum(\lambda'_x - \lambda_x)\widehat{l}_x} \tau_U(l) \; - \; 1)\mu_e(dl) \; = \; \log(\frac{\mathcal{Z}_{e',U}}{\mathcal{Z}_e})$.

(b) $\mathbb{E}(\prod_{x,y}[\frac{C'_{x,y}}{C_{x,y}}]^{N_{x,y}(\mathcal{L}_\alpha)} e^{-\sum(\lambda'_x - \lambda_x)\widehat{\mathcal{L}_\alpha}^x} \tau_U(\mathcal{L}_\alpha)) = (\frac{\mathcal{Z}_{e',U}}{\mathcal{Z}_e})^\alpha$.

Let us now introduce a new

Definition 3. We say that sets Λ_i of non-trivial loops are equivalent when the associated occupation fields are equal and when the total traversal numbers $\sum_{l \in \Lambda_i} N_{x,y}(l)$ are equal for all oriented edges (x,y). Equivalence classes will be called loop networks on the graph. We denote $\overline{\Lambda}$ the loop network defined by Λ.

Similarly, a set L of non-trivial discrete loops defines a discrete network characterized by the total traversal numbers.

The expectations computed in (6.7) determine the distribution of the network $\overline{\mathcal{L}_\alpha}$ defined by the loop ensemble $\mathcal{L}_\alpha$. We will denote $B^{e,e',\omega}(l)$ the variables

$$e^{\sum N_{x,y}(l) \log(\frac{C'_{x,y}}{C_{x,y}}) - \sum(\lambda'_x - \lambda_x)\widehat{l}_x + i\int_l \omega}$$

and $B^{e,e',\omega}(\mathcal{L}_\alpha)$ the variables

$$\prod_{l \in \mathcal{L}_\alpha} B^{e,e',\omega}(l) = \prod_{x,y}[\frac{C'_{x,y}}{C_{x,y}}e^{i\omega_{x,y}}]^{N_{x,y}(\mathcal{L}_\alpha)} e^{-\sum(\lambda'_x - \lambda_x)\widehat{\mathcal{L}_\alpha}^x}.$$

More generally, we can define $B^{e,e',U}(l)$ and $B^{e,e',U}(\mathcal{L}_\alpha)$ in a similar way as $B^{e,e',\omega}(l)$ and $B^{e,e',\omega}(\mathcal{L}_\alpha)$, using $\tau_U(l)$ instead of $e^{i\int_l \omega}$. Note that for each fixed e, when U and e' vary with $\frac{C'}{C} \leq 1$ and $\lambda' = \lambda$, linear combinations of the variables $B^{e,e',U}(\mathcal{L}_\alpha)$ form an algebra as $B^{e,e'_1,U_1}B^{e,e'_2,U_2} = B^{e,e'_{1,2},U_1 \otimes U_2}$, with $C^{e'_{1,2}} = \frac{C^{e'_1}C^{e'_2}}{C}$. In particular, $B^{e,e'_1,\omega_1}B^{e,e'_2,\omega_2} = B^{e,e'_{1,2},\omega_1+\omega_2}$.

Remark 15. Note that the expectations of the variables $B^{e,e',\omega}(\mathcal{L}_\alpha)$ determine the law of the network $\overline{\mathcal{L}}_\alpha$ defined by the loop ensemble $\mathcal{L}_\alpha$.

To work with μ, we should rather consider linear combinations of the form $\sum \lambda_i (B^{e,e'_i,U_i} - 1)$, with $\sum \lambda_i = 0$, which form also an algebra.

Remark 16. Formulas (6.6) and (6.7) apply to the calculation of loop indices: If we have for example a simple random walk on an oriented planar graph, and if z' is a point of the dual graph X', $\omega^{(z')}$ can be chosen such that for any loop l, $\int_l \omega^{(z')}$ is the winding number of the loop around a given point z' of the dual graph X'. Then $e^{i\pi \sum_{l \in \mathcal{L}_\alpha} \int_l \omega^{(z')}}$ is a spin system of interest. We then get for example that

$$\mu\left(\int_l \omega_{z'} \neq 0\right) = -\frac{1}{2\pi}\int_0^{2\pi} \log(\det(G^{(2\pi u \omega^{(z')})}G^{-1}))du$$

and hence

$$\mathbb{P}\left(\sum_{l \in \mathcal{L}_\alpha} |\int_l \omega^{(z')}| = 0\right) = e^{\frac{\alpha}{2\pi}\int_0^{2\pi} \log(\det(G^{(2\pi u \omega^{(z')})}G^{-1}))du}.$$

Conditional distributions of the occupation field with respect to values of the winding number can also be obtained.

Chapter 7
Decompositions

Note first that with the energy e, we can associate a time-rescaled Markov chain $\widehat{x}_t$ in which holding times at any point x are exponential times of parameters λ_x: $\widehat{x}_t = x_{\tau_t}$ with $\tau_t = \inf(s, \int_0^s \frac{1}{\lambda_{x_u}} du = t)$. For the time-rescaled Markov chain, local times coincide with the time spent in a point and the duality measure is simply the counting measure. The potential operator then essentially coincides with the Green function. The Markov loops can be time-rescaled as well and we did it in fact already when we introduced pointed loops. More generally we may introduce different holding time parameters but it would be rather useless as the random variables we are interested in are intrinsic, i.e. depend only on e.

7.1 Traces of Markov Chains and Energy Decomposition

If $D \subset X$ and we set $F = D^c$, the orthogonal decomposition of the energy $e(f, f) = e(f)$ into $e^D(f - H^F f) + e(H^F f)$ (see Proposition 17) leads to the decomposition of the Gaussian free field mentioned above and also to a decomposition of the time-rescaled Markov chain into the time-rescaled Markov chain killed at the exit of D and its trace on F, i.e. $\widehat{x}_t^{\{F\}} = \widehat{x}_{S_t^F}$, with $S_t^F = \inf(s, \int_0^s 1_F(\widehat{x}_u) du = t)$.

Proposition 24. *The trace of the time-rescaled Markov chain on F is the time-rescaled Markov chain defined by the energy functional $e^{\{F\}}(f) = e(H^F f)$, for which*

$$C_{x,y}^{\{F\}} = C_{x,y} + \sum_{a,b \in D} C_{x,a} C_{b,y} [G^D]^{a,b},$$

$$\lambda_x^{\{F\}} = \lambda_x - \sum_{a,b \in D} C_{x,a} C_{b,x} [G^D]^{a,b},$$

Y. Le Jan, *Markov Paths, Loops and Fields*, Lecture Notes in Mathematics 2026, DOI 10.1007/978-3-642-21216-1_7, © Springer-Verlag Berlin Heidelberg 2011

and

$$\mathcal{Z}_e = \mathcal{Z}_{eD}\,\mathcal{Z}_{e\{F\}}.$$

Proof. For the second assertion, note first that for any $y \in F$,

$$[H^F]_y^x = 1_{x=y} + 1_D(x)\sum_{b\in D}[G^D]^{x,b}C_{b,y}.$$

Moreover, $e(H^F f) = e(f, H^F f)$, by Proposition 17 and therefore

$$\lambda_x^{\{F\}} = e^{\{F\}}(1_{\{x\}}) = e(1_{\{x\}}, H^F 1_{\{x\}}) = \lambda_x - \sum_{a\in D} C_{x,a}[H^F]_x^a = \lambda_x(1 - p_x^{\{F\}})$$

where $p_x^{\{F\}} = \sum_{a,b\in D} P_a^x[G^D]^{a,b}C_{b,x} = \sum_{a\in D} P_a^x[H^F]_x^a$ is the probability that the Markov chain starting at x will first perform an excursion in D and then return to x.

Then for distinct x and y in F,

$$C_{x,y}^{\{F\}} = -e^{\{F\}}(1_{\{x\}}, 1_{\{y\}}) = -e(1_{\{x\}}, H^F 1_{\{y\}})$$

$$= C_{x,y} + \sum_a C_{x,a}[H^F]_y^a = C_{x,y} + \sum_{a,b\in D} C_{x,a}C_{b,y}[G^D]^{a,b}.$$

Note that the graph defined on F by the non-vanishing conductances $C_{x,y}^{\{F\}}$ has in general more edges than the restriction to F of the original graph.

For the third assertion, note also that $G^{\{F\}}$ is the restriction of G to F as for all $x, y \in F$, $e^{\{F\}}(G\delta_{y|F}, 1_{\{x\}}) = e(G\delta_y, [H^F 1_{\{x\}}]) = 1_{\{x=y\}}$. Hence the determinant decomposition already given in Sect. 4.3 yields the formula. The cases where F has one point was considered as a special case in Sect. 4.3.

For the first assertion note the transition matrix $[P^{\{F\}}]_y^x$ can be computed directly and equals

$$P_y^x + \sum_{a,b\in D} P_a^x[V^{D\cup\{x\}}]_b^a P_y^b = P_y^x + \sum_{a,b\in D} P_a^x[G^{D\cup\{x\}}]^{a,b}C_{b,y}.$$

It can be decomposed according to whether the jump to y occurs from x or from D and the number of excursions from x to x:

$$[P^{\{F\}}]_y^x = \sum_{k=0}^{\infty}(\sum_{a,b\in D} P_a^x[V^D]_b^a P_x^b)^k (P_y^x + \sum_{a,b\in D} P_a^x[V^D]_b^a P_y^b)$$

$$= \sum_{k=0}^{\infty}(\sum_{a,b\in D} P_a^x[G^D]^{a,b}C_{b,x})^k (P_y^x + \sum_{a,b\in D} P_a^x[G^D]^{a,b}C_{b,y}).$$

The expansion of $\dfrac{C_{x,y}^{\{F\}}}{\lambda_x^{\{F\}}}$ in geometric series yields exactly the same result.

Finally, remark that the holding times of $\widehat{x}_t^{\{F\}}$ at any point $x \in F$ are sums of a random number of independent holding times of $\widehat{x}_t$. This random integer counts the excursions from x to x performed by the chain $\widehat{x}_t$ during the holding time of $\widehat{x}_t^{\{F\}}$. It follows a geometric distribution of parameter $1 - p_x^{\{F\}}$. Therefore, $\frac{1}{\lambda_x^{\{F\}}} = \frac{1}{\lambda_x(1-p_x^{\{F\}})}$ is the expectation of the holding times of $\widehat{x}_t^{\{F\}}$ at x. $\qquad\square$

7.2 Excursion Theory

A loop in X which hits F can be decomposed into a loop in F and its excursions in D which may come back to their starting point.

More precisely, a loop l hitting F can be decomposed into its restriction $l^{\{F\}} = (\xi_i, \widehat{\tau}_i)$ in F (possibly a one point loop), a family of excursions $\gamma_{\xi_i, \xi_{i+1}}$ attached to the jumps of $l^{\{F\}}$ and systems of i.i.d. excursions $(\gamma_{\xi_i}^h, h \le n_{\xi_i})$ attached to the points of $l^{\{F\}}$. These sets of excursions can be empty.

Let $\mu_D^{a,b}$ denote the bridge measure (with mass $[G^D]^{a,b}$) associated with e^D. Set

$$\nu_{x,y}^D = \frac{1}{C_{x,y}^{\{F\}}}[C_{x,y}\delta_\emptyset + \sum_{a,b\in D} C_{x,a}C_{b,y}\mu_D^{a,b}], \quad \nu_x^D = \frac{1}{\lambda_x p_x^{\{F\}}}(\sum_{a,b\in D} C_{x,a}C_{b,x}\mu_D^{a,b})$$

and note that $\nu_{x,y}^D(1) = \nu_x^D(1) = 1$.

Let μ^D be the restriction of μ to loops in contained in D. It is the loop measure associated to the process killed at the exit of D. We get a decomposition of $\mu - \mu^D$ in terms of the loop measure $\mu^{\{F\}}$ defined on loops of F by the trace of the Markov chain on F, probability measures $\nu_{x,y}^D$ on excursions in D indexed by pairs of points in F and ν_x^D on excursions in D indexed by points of F. Moreover, conditionally on $l^{\{F\}}$, the integers n_{ξ_i} follow a Poisson distribution of parameter $(\lambda_{\xi_i} - \lambda_{\xi_i}^{\{F\}})\widehat{\tau}_i$ (the total holding time in ξ_i before another point of F is visited) and the conditional distribution of the rescaled holding times in ξ_i before each excursion $\gamma_{\xi_i}^l$ is the distribution $\beta_{n_{\xi_i}, \widehat{\tau}_i}$ of the increments of a uniform sample of n_{ξ_i} points in $[0\ \widehat{\tau}_i]$ put in increasing order. We denote these holding times by $\widehat{\tau}_{i,h}$ and set $l = \Lambda(l^{\{F\}}, (\gamma_{\xi_i, \xi_{i+1}}), (n_{\xi_i}, \gamma_{\xi_i}^h, \widehat{\tau}_{i,h}))$.

Then $\mu - \mu^D$ is the image measure by Λ of

$$\mu^{\{F\}}(dl^{\{F\}}) \prod(\nu_{\xi_i, \xi_{i+1}}^D)(d\gamma_{\xi_i, \xi_{i+1}})e^{-(\lambda_{\xi_i} - \lambda_{\xi_i}^{\{F\}})\widehat{\tau}_i}$$

$$\sum_k \frac{[(\lambda_{\xi_i} - \lambda_{\xi_i}^{\{F\}})\widehat{\tau}_i]^k}{k!}\delta_{n_{\xi_i}}^k[\nu_x^D]^{\otimes k}(d\gamma_{\xi_i}^h)\beta_{k,\widehat{\tau}_i}(d\widehat{\tau}_{i,h}).$$

Note that for x, y belonging to F, the bridge measure $\mu^{x,y}$ can be decomposed in the same way, with the same excursion measures.

7.2.1   The One Point Case and the Excursion Measure

If F is reduced to a point x_0, and κ vanishes on $D = \{x_0\}^c$, the decomposition is of course simpler.

First, $\lambda_{x_0} = \sum_a C_{x_0,a} + \kappa_x$ and $\lambda_{x_0}^{\{x_0\}} = \kappa_{x_0}$. Then,

$$p_{x_0}^{\{x_0\}} = \sum_{a,b \in D} P_{x_0}^{x_0}[G^D]^{a,b} C_{b,x_0} = \frac{\sum C_{x_0,a}}{\lambda_{x_0}} = 1 - \frac{\kappa_{x_0}}{\lambda_{x_0}},$$

as $C_{\cdot,x_0}$ is the killing measure of e^D and therefore its G^D potential equals 1. $l^{\{x_0\}}$ is a trivial one point loop with rescaled lifetime $\widehat{\tau} = \widehat{l}^{x_0} \frac{\lambda_{x_0}}{\kappa_{x_0}} = \frac{\widehat{l}^{x_0}}{1 - p_{x_0}^{\{x_0\}}}$ and the number of excursions (all independent with the same distribution $\rho_{x_0}^{\{x_0\}^c}$) follows a Poisson distribution of parameter $\kappa_{x_0} \widehat{\tau} = \lambda_{x_0} \widehat{l}^{x_0}$.

The non-normalized excursion measure

$$\rho^D = (\lambda_{x_0} - \kappa_{x_0}) \nu_{x_0}^D = \sum_{a,b \in D} C_{x_0,a} C_{b,x_0} \mu_D^{a,b}$$

verifies the following property: for any subset K of D,

$$\rho^D(\{\gamma, \widehat{\gamma}(K) > 0\}) = Cap_{e^D}(K).$$

Indeed, the lefthand side can be expressed as

$$\sum_{a,b \in D} C_{x_0,a} C_{b,x_0} [H^K G^D]^{a,b} = \sum_{a \in D} C_{x_0,a} [H^K 1]^a = e^D(1, H^K 1).$$

It should be noted that ρ^D depends only of e^D (i.e. does not depend on κ_{x_0}).

Proposition 25. *(a) Under ρ^D, the non-normalized hitting distribution of any $K \subseteq D$ is the e^D-capacitary measure of K. The same property holds for the last hitting distribution.*

(b) Under $\rho^D(d\gamma)$, the conditional distribution of the path γ between $T_K(\gamma)$ (the first time in K) and $L_K(\gamma)$ (the last time in K), given γ_{T_K} and γ_{L_K} is $\frac{1}{[G^D]^{\gamma_{T_K}, \gamma_{L_K}}} \mu_D^{\gamma_{T_K}, \gamma_{L_K}}$

Proof. (a) By definition of ρ^D, the non-normalized hitting distribution of K is expressed for any $z \in K$ by $\sum_{a,b,c \in D} C_{x_0,a} C_{b,x_0} [G^{D-K}]^{a,c} C_{c,z} [G^D]^{z,b} = \sum_{a,c \in D} C_{x_0,a} [G^{D-K}]^{a,c} C_{c,z}$. $C_{x_0,a}, a \in D$ is the killing measure of e^D and $[G^{D-K}]^{a,c} C_{c,z}$ the e^D-balayage kernel on K. The case of last hitting distribution follows from the invariance of ρ^D under time reversal.

(b) Indeed, on functions of a path after T_K,

$$\mu_D^{a,b} = \sum_{a,c \in D, z \in K} C_{x_0,a} [G^{D-K}]^{a,c} C_{c,z} \mu_D^{z,b}$$

and therefore on functions of a path restricted to $[T_K, L_K]$, ρ^D equals:

$$\sum_{z \in K} \sum_{a,b,c \in D} C_{x_0,a} [G^{D-K}]^{a,c} C_{c,z} \mu_D^{z,b} C_{b,x_0}$$

$$= \sum_{z,t \in K} \sum_{a,b,c,d \in D} C_{x_0,a} [G^{D-K}]^{a,c} C_{c,z} \mu_D^{z,t} C_{t,d} [G^{D-K}]^{d,b} C_{b,x_0}. \quad \square$$

Remark 17. This construction of ρ^D can be extended to transient chains on infinite spaces with zero killing measure. There exists a unique measure on equivalence classes under the shift of doubly infinite paths converging to infinity on both sides, such that the hitting distribution of any compact set is given by its capacitary measure (Cf. [12, 46, 56], and the first section of [53] for a recent presentation in the case of $\mathbb{Z}^d$ random walks). Proposition 25 holds also in this context.

Following [53], the set of points hit by a Poissonian set of excursions of intensity $\alpha \rho^D$ can be called the interlacement at level α.

The law $\frac{\mu^{x_0,x_0}}{G^{x_0,x_0}}$ can of course be decomposed in the same way, with the same conditional distribution given $\widehat{l}^{x_0}$. Recall that by Proposition 9, $\widehat{l}^{x_0}$ follows an exponential distribution with mean G^{x_0,x_0}.

$\widehat{\mathcal{L}}_\alpha^{x_0}$ follows a $\Gamma(\alpha, G^{x_0,x_0})$ distribution, in particular an exponential distribution with mean G^{x_0,x_0} for $\alpha = 1$. Moreover, the union of the excursions of all loops of $\mathcal{L}_\alpha$ outside x_0 has obviously the same Poissonian conditional distribution, given $\widehat{\mathcal{L}}_\alpha^{x_0} = s$ than μ and $\frac{\mu^{x_0,x_0}}{G^{x,x}}$, given $\widehat{l}^{x_0} = s$. The set of excursions outside x_0 defined by the $\frac{\mu^{x_0,x_0}}{G^{x_0,x_0}}$-distributed bridge and by $\mathcal{L}_1$ are therefore identically distributed, as the total holding time in x_0.

Remark 18. Note finally that by Exercise 13, the distribution of $\mathcal{L}_1/\mathcal{L}_1^D$ can be recovered from a unique sample of $\frac{\mu^{x_0,x_0}}{G^{x_0,x_0}}$ by splitting the bridge according to an independent sample U_i of $Poisson - Dirichlet(0, \alpha)$, more precisely, by splitting the bridge (in fact a based loop) l into based subloops $l|_{[\sigma_i, \sigma_{i+1}]}$, with $\sigma_i = \inf(s, \frac{1}{\lambda_x} \int_0^s 1_{\{x_0\}}(l_s) ds = \sum_1^i U_j \widehat{l}^{x_0})$.

Conversely, a sample of the bridge could be recovered from a sample of the loop set $\mathcal{L}_1/\mathcal{L}_1^D$ by concatenation in random order. This random ordering can be defined by taking a projective limit of the randomly ordered finite subset of loops $\{l_{i,n}\}$ defined by assuming for example that $\widehat{l}_{i,n}^x > \frac{1}{n}$.

7.3   Conditional Expectations

Coming back to the general case, the Poisson process $\mathcal{L}_\alpha^{\{F\}} = \{l^{\{F\}}, l \in \mathcal{L}_\alpha\}$ has intensity $\mu^{\{F\}}$ and is independent of $\mathcal{L}_\alpha^D$.

Note that $\widehat{\mathcal{L}_\alpha^{\{F\}}}$ is the restriction of $\widehat{\mathcal{L}_\alpha}$ to F.

If χ is carried by D and if we set $e_\chi = e + \|\quad\|_{L^2(\chi)}$ and denote $[e_\chi]^{\{F\}}$ by $e^{\{F,\chi\}}$ we have

$$C_{x,y}^{\{F,\chi\}} = C_{x,y} + \sum_{a,b} C_{x,a} C_{b,y} [G_\chi^D]^{a,b}, \quad p_x^{\{F,\chi\}} = \sum_{a,b\in D} P_a^x [G_\chi^D]^{a,b} C_{b,x}$$

and $\lambda_x^{\{F,\chi\}} = \lambda_x(1 - p_x^{\{F,\chi\}})$.

More generally, if $e^\#$ is such that $C^\# = C$ on $F \times F$, and $\lambda = \lambda^\#$ on F we have:

$$C_{x,y}^{\#\{F\}} = C_{x,y} + \sum_{a,b} C_{x,a}^\# C_{b,y}^\# [G^{\#D}]^{a,b}, \quad p_x^{\#\{F\}} = \sum_{a,b\in D} P_a^{\#x} [G^{\#D}]^{a,b} C_{b,x}^\#$$

and $\lambda_x^{\#\{F\}} = \lambda_x(1 - p_x^{\#\{F\}})$.

If χ is a measure carried by D, we have:

$$\mathbb{E}(e^{-\langle \widehat{\mathcal{L}_\alpha}, \chi\rangle} | \mathcal{L}_\alpha^{\{F\}}) = \mathbb{E}(e^{-\langle \widehat{\mathcal{L}_\alpha^D}, \chi\rangle})(\prod_{x,y\in F} [\int e^{-\langle\widehat{\gamma},\chi\rangle} \nu_{x,y}^D(d\gamma)]^{N_{x,y}(\mathcal{L}_\alpha^{\{F\}})}$$

$$\times \prod_{x\in F} e^{(\lambda_x - \lambda_x^{\{F\}})[\widehat{\mathcal{L}_\alpha^{\{F\}}}]^x \int (e^{-\langle\widehat{\gamma},\chi\rangle}-1)\nu_x^D(d\gamma)}$$

$$= [\frac{\mathcal{Z}_{e_\chi^D}}{\mathcal{Z}_{e^D}}]^\alpha (\prod_{x,y\in F} [\frac{C_{x,y}^{\{F,\chi\}}}{C_{x,y}^{\{F\}}}]^{N_{x,y}(\mathcal{L}_\alpha^{\{F\}})} \prod_{x\in F} e^{[-\lambda_x^{\{F,\chi\}}+\lambda_x^{\{F\}}]\widehat{\mathcal{L}_\alpha^x}}.$$

(recall that $\widehat{\mathcal{L}_\alpha^{\{F\}}}$ is the restriction of $\widehat{\mathcal{L}_\alpha}$ to F). Also, if we condition on the set of discrete loops $\mathcal{DL}_\alpha^{\{F\}}$

$$\mathbb{E}(e^{-\langle \widehat{\mathcal{L}_\alpha}, \chi\rangle} | \mathcal{DL}_\alpha^{\{F\}}) = [\frac{\mathcal{Z}_{e_\chi^D}}{\mathcal{Z}_{e^D}}]^\alpha (\prod_{x,y\in F} [\frac{C_{x,y}^{\{F,\chi\}}}{C_{x,y}^{\{F\}}}]^{N_{x,y}(\mathcal{L}_\alpha^{\{F\}})} \prod_{x\in F} [\frac{\lambda_x^{\{F\}}}{\lambda_x^{\{F,\chi\}}}]^{N_x(\mathcal{L}_\alpha^{\{F\}})+\alpha})$$

where the last exponent $N_x + 1$ is obtained by taking into account the loops which have a trivial trace on F (see formula (6.2)).

More generally we can show in the same way the following.

Proposition 26. *If $C^\# = C$ on $F \times F$, and $\lambda = \lambda^\#$ on F, we denote $B^{e,e^\#}$ the multiplicative functional*

$$\prod_{x,y} [\frac{C^\#_{x,y}}{C_{x,y}}]^{N_{x,y}} e^{-\sum_{x \in D} \hat{l}_x (\lambda^\#_x - \lambda_x)}.$$

Then,

$$\mathbb{E}(B^{e,e^\#} | \mathcal{L}^{\{F\}}_\alpha) = [\frac{\mathcal{Z}_{e^\#D}}{\mathcal{Z}_{e^D}}]^\alpha \prod_{x,y \in F} [\frac{C^{\#\{F\}}_{x,y}}{C^{\{F\}}_{x,y}}]^{N_{x,y}(\mathcal{L}^{\{F\}}_\alpha)} \prod_{x \in F} e^{[-\lambda^{\#\{F\}}_x + \lambda^{\{F\}}_x] \widehat{\mathcal{L}^x_\alpha}}$$

and

$$\mathbb{E}(B^{e,e^\#} | \mathcal{DL}^{\{F\}}_\alpha) = [\frac{\mathcal{Z}_{e^\#D}}{\mathcal{Z}_{e^D}}]^\alpha \prod_{x,y \in F} [\frac{C^{\#\{F\}}_{x,y}}{C^{\{F\}}_{x,y}}]^{N_{x,y}(\mathcal{L}^{\{F\}}_\alpha)} \prod_{x \in F} [\frac{\lambda^{\{F\}}_x}{\lambda^{\#\{F\}}_x}]^{N_x(\mathcal{L}^{\{F\}}_\alpha) + \alpha}.$$

These decomposition and conditional expectation formulas extend to include a current ω in $C^\#$. Note that if ω is closed (i.e. vanish on every loop) in D, one can define ω^F such that $[Ce^{i\omega}]^{\{F\}} = C^{\{F\}} e^{i\omega^F}$. Then

$$\mathcal{Z}_{e,\omega} = \mathcal{Z}_{e^D} \mathcal{Z}_{e^{\{F\}}, \omega^F}.$$

The previous proposition implies the following *Markov property*:

Remark 19. If $D = D_1 \cup D_2$ with D_1 and D_2 strongly disconnected, (i.e. such that for any $(x, y, z) \in D_1 \times D_2 \times F$, $C_{x,y}$ and $C_{x,z} C_{y,z}$ vanish), the restrictions of the network $\overline{\mathcal{L}_\alpha}$ to $D_1 \cup F$ and $D_2 \cup F$ are independent conditionally on the restriction of $\mathcal{L}_\alpha$ to F.

Proof. This follows from the fact that as D_1 and D_2 are strongly disconnected, any excursion measure $\nu^D_{x,y}$ or ρ^D_x from F into $D = D_1 \cup D_2$ is an excursion measure either in D_1 or in D_2. $\square$

7.4 Branching Processes with Immigration

An interesting example can be given after extending slightly the scope of the theory to countable transient symmetric Markov chains: We can take $X = \mathbb{N} - \{0\}$, $C_{n,n+1} = 1$ for all $n \geq 1$, $\kappa_n = 0$ for $n \geq 2$ and $\kappa_1 = 1$. P is the transfer matrix of the simple symmetric random walk killed at 0.

Then we can apply the previous considerations to check that $\widehat{\mathcal{L}}_\alpha^n$ is a branching process with immigration.

The immigration at level n comes from the loops whose infimum is n and the branching from the excursions to level $n+1$ of the loops existing at level n. Set $F_n = \{1, 2, ..., n\}$ and $D_n = F_n^c$.

From the calculations of conditional expectations made above, we get that for any positive parameter γ,

$$\mathbb{E}(e^{-\gamma\widehat{\mathcal{L}}_\alpha^n}||\mathcal{L}_\alpha^{\{F_{n-1}\}}) = \mathbb{E}(e^{-\gamma[\widehat{\mathcal{L}_\alpha^{D_{n-1}}}]^n})e^{[\lambda_{n-1}^{\{F_{n-1},\gamma\delta_n\}} - \lambda_{n-1}^{\{F_{n-1}\}}]\widehat{\mathcal{L}}_\alpha^{n-1}}$$

($[\widehat{\mathcal{L}_\alpha^{D_{n-1}}}]^n$ denotes the occupation field of the trace of $\mathcal{L}_\alpha$ on D_{n-1} evaluated at n).

From this formula, it is clear that $\widehat{\mathcal{L}}_\alpha^n$ is a branching Markov chain with immigration. To be more precise, note that for any $n, m > 0$, the potential operator V_m^n equals $2(n \wedge m)$ that $\lambda_n = 2$ and that $G^{1,1} = 1$. Moreover, by the generalized resolvent equation, $G_{\gamma\delta_1}^{1,n} = G^{1,n} - G^{1,1}\gamma G_{\gamma\delta_1}^{1,n}$ so that $G_{\gamma\delta_1}^{1,n} = \frac{1}{1+\gamma}$. For any $n > 0$, the restriction of the Markov chain to D_n is isomorphic to the original Markov chain. Then it comes that for all n, $p_n^{\{F_n\}} = \frac{1}{2}$, $\lambda_n^{\{F_n\}} = 1$, and $\lambda_n^{\{F_n,\gamma\delta_{n+1}\}} = 2 - \frac{1}{1+\gamma} = \frac{2\gamma+1}{1+\gamma}$ so that the Laplace exponent of the convolution semigroup ν_t defining the branching mechanism $\lambda_{n-1}^{\{F_{n-1},\gamma\delta_n\}} - \lambda_{n-1}^{\{F_{n-1}\}}$ equals $\frac{2\gamma+1}{1+\gamma} - 1 = \frac{\gamma}{1+\gamma} = \int(1 - e^{-\gamma s})e^{-s}ds$. It is the semigroup of a compound Poisson process whose Levy measure is exponential.

The immigration law (on $\mathbb{R}^+$) is a Gamma distribution $\Gamma(\alpha, G^{1,1}) = \Gamma(\alpha, 1)$. It is the law of $\widehat{\mathcal{L}}_\alpha^1$ and also of $[\widehat{\mathcal{L}_\alpha^{D_{n-1}}}]^n$ for all $n > 1$.

The conditional law of $\widehat{\mathcal{L}}_\alpha^{n+1}$ given $\widehat{\mathcal{L}}_\alpha^n$ is the convolution of the immigration law $\Gamma(\alpha, 1)$ with $\nu_{\widehat{\mathcal{L}}_\alpha^n}$

Exercise 29. Alternatively, we can consider the integer valed process $N_n(\mathcal{L}_\alpha^{\{F_n\}}) + 1$ which is a Galton Watson process with immigration. In our example, we find the reproduction law $\pi(n) = 2^{-n-1}$ for all $n \geq 0$ (critical binary branching).

Exercise 30. Show that more generally, if $C_{n,n+1} = [\frac{p}{1-p}]^n$, for $n > 0$ and $\kappa_1 = 1$, with $0 < p < 1$, we get all asymetric simple random walks. Show that $\lambda_n = \frac{p^{n-1}}{(1-p)^n}$ and $G^{1,1} = 1$. Determine the distributions of the associated branching and Galton Watson process with immigration.

If we consider the occupation field defined by the loops whose infimum equals 1 (i.e. going through 1), we get a branching process without immigration: it is the classical relation between random walks local times and branching processes.

7.5 Another Expression for Loop Hitting Distributions

Let us come back to formula (4.9). Setting $F = F_1 \cup F_2$, we see that this result involves only $\mu^{\{F\}}$ and $e^{\{F\}}$ i.e. it can be expressed interms of the restrictions of the loops to F.

Lemma 1. *If* $X = X_1 \cup X_2$ *with* $X_1 \cap X_2 = \emptyset$,

$$\log\Big(\frac{\det(G)}{\det(G^{X_1})\det(G^{X_2})}\Big) = \sum_1^\infty \frac{1}{2k} Tr([H_{12}H_{21}]^k + [H_{12}H_{21}]^k)$$

with $H_{12} = H^{X_2}|_{X_1}$ *and* $H_{21} = H^{X_1}|_{X_2}$.

Proof.

$$\frac{\det(G)}{\det(G^{X_1})\det(G^{X_2})} = \Big(\det\Big(\begin{matrix} I_{X_1 \times X_1} & -G^{X_1}C_{X_1 \times X_2} \\ -G^{X_2}C_{X_2 \times X_1} & I_{X_2 \times X_2} \end{matrix}\Big)\Big)^{-1}$$

$$= \Big(\det\Big(\begin{matrix} I_{X_1 \times X_1} & -H_{12} \\ -H_{21} & I_{X_2 \times X_2} \end{matrix}\Big)\Big)^{-1}.$$

The transience implies that either $H_{12}1$, *either* $H_{21}1$ *is strictly less than 1, and therefore,* $H_{12}H_{21}$ *and* $H_{12}H_{21}$ *are strict contractions. From the expansion of* $-\log(1-x)$, *we get that:*

$$\log\Big(\frac{\det(G)}{\det(G^{X_1})\det(G^{X_2})}\Big) = \sum_1^\infty \frac{1}{k} Tr\Big[\Big(\begin{matrix} 0 & -H_{12} \\ -H_{21} & 0 \end{matrix}\Big)^k\Big].$$

The result follows, as odd terms have obviously zero trace. □

Noting finally that the lemma can be applied to the restrictions of G to $F_1 \cup F_2$, F_1 and F_2, and that hitting distributions of F_1 from F_2 and F_2 from F_1 are the same for the Markov chain on X and its restriction to $F_1 \cup F_2$, we get finally:

Proposition 27. *If* F_1 *and* F_2 *are disjoint,*

$$\mu(\widehat{l}(F_1)\widehat{l}(F_2) > 0) = \sum_1^\infty \frac{1}{2k} Tr([H_{12}H_{21}]^k + [H_{21}H_{12}]^k)$$

with $H_{12} = H^{F_2}|_{F_1}$ *and* $H_{21} = H^{F_1}|_{F_2}$.

Exercise 31. Show that the k-th term of the expansion can be interpreted as the measure of loops with exactly k-crossings between F_1 and F_2.

Exercise 32. Prove analogous results for n disjoint sets F_i.

Chapter 8
Loop Erasure and Spanning Trees

8.1 Loop Erasure

Recall that an oriented link g is a pair of points (g^-, g^+) such that $C_g = C_{g^-, g^+} \neq 0$. Define $-g = (g^+, g^-)$.

Let $\mu_{\neq}^{x,y}$ be the measure induced by C on discrete self-avoiding paths between x and y: $\mu_{\neq}^{x,y}(x, x_2, ..., x_{n-1}, y) = C_{x,x_2} C_{x_1,x_3} ... C_{x_{n-1},y}$.

Another way to define a measure on discrete self avoiding paths from x to y from a measure on paths from x to y is loop erasure defined in Sect. 3.1 (see also [16, 17, 39] and [31]). In this context, the loops, which can be reduced to points, include holding times, and loop erasure produces a discrete path without holding times.

We have the following:

Theorem 4. *The image of $\mu^{x,y}$ by the loop erasure map $\gamma \to \gamma^{BE}$ is $\mu_{BE}^{x,y}$ defined on self avoiding paths by*

$$\mu_{BE}^{x,y}(\eta) = \mu_{\neq}^{x,y}(\eta) \frac{\det(G)}{\det(G^{\{\eta\}^c})} = \mu_{\neq}^{x,y}(\eta) \det(G_{|\{\eta\} \times \{\eta\}})$$

(Here $\{\eta\}$ denotes the set of points in the path η) and by $\mu_{BE}^{x,y}(\emptyset) = \delta_y^x G_{x,x}$.

Proof. Set $\eta = (x_1 = x, x_2, ..., x_n = y)$ and $\eta_m = (x, ..., x_m)$, for any $m > 1$. Then,

$$\mu^{x,y}(\gamma^{BE} = \eta) = \sum_{k=0}^{\infty} [P^k]_x^x P_{x_2}^x \mu_{\{x\}^c}^{x_2,y}(\gamma^{BE} = \theta\eta) = V_x^x P_{x_2}^x \mu_{\{x\}^c}^{x_2,y}(\gamma^{BE} = \theta\eta)$$

where $\mu_{\{x\}^c}^{x_2,y}$ denotes the bridge measure for the Markov chain killed as it hits x and θ the natural shift on discrete paths. By recurrence, this clearly equals

$$V_x^x P_{x_2}^x [V^{\{x\}^c}]_{x_2}^{x_2} ... [V^{\{\eta_{n-1}\}^c}]_{x_{n-1}}^{x_{n-1}} P_y^{x_{n-1}} [V^{\{\eta\}^c}]_y^y \lambda_y^{-1} = \mu_{\neq}^{x,y}(\eta) \frac{\det(G)}{\det(G^{\{\eta\}^c})}$$

Y. Le Jan, *Markov Paths, Loops and Fields*, Lecture Notes in Mathematics 2026, DOI 10.1007/978-3-642-21216-1_8, © Springer-Verlag Berlin Heidelberg 2011

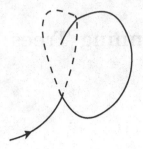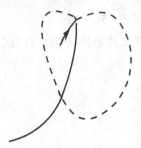

Fig. 8.1 Orientation dependance

as

$$[V^{\{\eta_{m-1}\}^c}]_{x_m}^{x_m} = \frac{\det([[I-P]|_{\{\eta_m\}^c \times \{\eta_m\}^c})}{\det([[I-P]|_{\{\eta_{m-1}\}^c \times \{\eta_{m-1}\}^c})} = \frac{\det(V^{\{\eta_{m-1}\}^c})}{\det(V^{\{\eta_m\}^c})}$$

$$= \frac{\det(G^{\{\eta_{m-1}\}^c})}{\det(G^{\{\eta_m\}^c})} \lambda_{x_m}.$$

for all $m \le n-1$. $\qquad\qquad\square$

Remark 20. It is worth noticing that, also the operation of loop erasure clearly depends on the orientation of the path (as shown in Fig. 8.1), the distribution of the loop erased bridge is reversible.

Also, by Feynman–Kac formula, for any self-avoiding path η:

$$\int e^{-<\hat{\gamma}, \chi>} 1_{\{\gamma^{BE}=\eta\}} \mu^{x,y}(d\gamma) = \frac{\det(G_\chi)}{\det(G_\chi^{\{\eta\}^c})} \mu_{\ne}^{x,y}(\eta) = \det(G_\chi)_{|\{\eta\} \times \{\eta\}} \mu_{\ne}^{x,y}(\eta)$$

$$= \frac{\det(G_\chi)_{|\{\eta\} \times \{\eta\}}}{\det(G_{|\{\eta\} \times \{\eta\}})} \mu_{BE}^{x,y}(\eta).$$

Therefore, recalling that by the results of Sect. 4.3 conditionally on η, $\mathcal{L}_1/\mathcal{L}_1^{\{\eta\}^c}$ and $\mathcal{L}_1^{\{\eta\}^c}$ are independent, we see that under $\mu^{x,y}$, the conditional distribution of $\hat{\gamma}$ given $\gamma^{BE} = \eta$ is the distribution of $\hat{\mathcal{L}}_1 - \hat{\mathcal{L}}_1^{\{\eta\}^c}$ i.e. the occupation field of the loops of $\mathcal{L}_1$ which intersect η.

More generally, it can be shown that

Proposition 28. *The conditional distribution of the network $\overline{\mathcal{L}_\gamma}$ defined by the loops of γ, given that $\gamma^{BE} = \eta$, is identical to the distribution of the network defined by $\mathcal{L}_1/\mathcal{L}_1^{\{\eta\}^c}$ i.e. the loops of $\mathcal{L}_1$ which intersect η.*

Proof. Recall the notation $\mathcal{Z}_e = \det(G)$. First an elementary calculation using (2.8) shows that $\mu_{e'}^{x,y}(e^{i\int_\gamma \omega} 1_{\{\gamma^{BE}=\eta\}})$ equals

$$\mu_e^{x,y}\Big(1_{\{\gamma^{BE}=\eta\}}\prod[\frac{C'_{\xi_i,\xi_{i+1}}}{C_{\xi_i,\xi_{i+1}}}e^{i\omega_{\xi_i,\xi_{i+1}}}\frac{\lambda_{\xi_i}}{\lambda'_{\xi_i}}]\Big)$$

$$\frac{C'_{x,x_2}C'_{x_1,x_3}\dots C'_{x_{n-1},y}}{C_{x,x_2}C_{x_1,x_3}\dots C_{x_{n-1},y}}e^{i\int_\eta \omega}\mu_e^{x,y}\Big(\prod_{u\neq v}[\frac{C'_{u,v}}{C_{u,v}}e^{i\omega_{u,v}}]^{N_{u,v}(\mathcal{L}_\gamma)}e^{-\langle\lambda'-\lambda,\widehat{\gamma}\rangle}1_{\{\gamma^{BE}=\eta\}}\Big).$$

(Note the term $e^{-\langle\lambda'-\lambda,\widehat{\gamma}\rangle}$ can be replaced by $\prod_u(\frac{\lambda_u}{\lambda'_u})^{N_u(\gamma)+1}$.)

Moreover, by the proof of the previous proposition, applied to the Markov chain defined by e' perturbed by ω, we have also

$$\mu_{e'}^{x,y}(e^{i\int_\gamma \omega}1_{\{\gamma^{BE}=\eta\}}) = C'_{x,x_2}C'_{x_1,x_3}\dots C'_{x_{n-1},y}e^{i\int_\eta \omega}\frac{\mathcal{Z}_{e',\omega}}{\mathcal{Z}_{[e']\{\eta\}^c,\omega}}.$$

Therefore,

$$\mu_e^{x,y}(\prod_{u\neq v}[\frac{C'_{u,v}}{C_{u,v}}e^{i\omega_{u,v}}]^{N_{u,v}(\mathcal{L}_\gamma)}e^{-\langle\lambda'-\lambda,\widehat{\gamma}\rangle}|\gamma^{BE}=\eta) = \frac{\mathcal{Z}_{e\{\eta\}^c}\mathcal{Z}_{e',\omega}}{\mathcal{Z}_e\mathcal{Z}_{[e']\{\eta\}^c,\omega}}.$$

Moreover, by (6.7) and the properties of the Poisson processes,

$$\mathbb{E}(\prod_{u\neq v}[\frac{C'_{u,v}}{C_{u,v}}e^{i\omega_{u,v}}]^{N_{u,v}(\mathcal{L}_1/\mathcal{L}_1^{\{\eta\}^c})}e^{-\langle\lambda'-\lambda,\widehat{\mathcal{L}}_1-\widehat{\mathcal{L}}_1^{\{\eta\}^c}\rangle}) = \frac{\mathcal{Z}_{e\{\eta\}^c}\mathcal{Z}_{e',\omega}}{\mathcal{Z}_e\mathcal{Z}_{[e']\{\eta\}^c,\omega}}.$$

It follows that the joint distribution of the traversal numbers and the occupation field are identical for the set of erased loops and $\mathcal{L}_1/\mathcal{L}_1^{\{\eta\}^c}$. □

The general study of loop erasure which is done in this chapter yields the following result when applied to a universal covering $\widehat{X}$. Let $\widehat{G}$ be the Green function associated with the lift of the Markov chain.

Corollary 5. *The image of $\mu^{x,y}$ under the reduction map is given as follows: If c is a geodesic arc between x and y: $\mu^{x,y}(\{\xi, \xi^R = c\}) = \prod C_{c_i,c_{i+1}}\det(\widehat{G}_{|\{c\}\times\{c\}})$.*

Besides, if $\widehat{x}$ and $\widehat{y}$ are the endpoints of the lift of c to a universal covering,

$$\mu^{x,y}(\{\xi, \xi^R = c\}) = \widehat{G}_{\widehat{x},\widehat{y}}.$$

Note this yields an interesting identity on the Green function $\widehat{G}$.

Exercise 33. Check it in the special case treated in Proposition 13.

Similarly one can define the image of $\mathbb{P}^x$ by BE and check it is given by

$$\mathbb{P}^x_{BE}(\eta) = \delta^x_{x_1} C_{x_1,x_2}...C_{x_{n-1},x_n} \kappa_{x_n} \det(G_{|\{\eta\}-\Delta \times \{\eta\}-\Delta})$$

$$= \delta^x_{x_1} C_{x_1,x_2}...C_{x_{n-1},x_n} \kappa_{x_n} \frac{\det(G)}{\det(G^{\{\eta\}^c})}$$

for $\eta = (x_1, ..., x_n, \Delta)$.

Note that in particular, $\mathbb{P}^x_{BE}((x,\Delta)) = V^x_x(1 - \sum_y P^x_y) = \kappa_x G^{x,x}$.

Slightly more generally, one can determine the law of the image, by loop erasure path killed at it hits a subset F, the hitting point being now the end point of the loop erased path (instead of Δ, unless F is not hit during the lifetime of the path). If $x \in D = F^c$ is the starting point and $y \in F^\Delta$, the probability of $\eta = (x_1, ..., x_n, y)$ is

$$\delta^x_{x_1} C_{x_1,x_2}...C_{x_{n-1},x_n} C_{x_n,y} \det(G^D_{|\{\eta\}-y\times\{\eta\}-y}) = \delta^x_{x_1} C_{x_1,x_2}...C_{x_n,y} \frac{\det(G^D)}{\det(G^{D-\{\eta\}})}.$$

8.2 Wilson Algorithm

Wilson's algorithm (see [29]) iterates this last construction, starting with the points x arranged in an arbitrary order. The first step of the algorithm is the construction of a loop erased path starting at the first point and ending at Δ. This loop erased path is the first branch of the spanning tree. Each step of the algorithm reproduces this first step except it starts at the first point which is not visited by the already constructed tree of self avoiding paths, and stops when it hits that tree, or Δ, producing a new branch of the tree. This algorithm provides a construction, branch by branch, of a random spanning tree rooted in Δ. It turns out, as we will show below, that the distribution of this spanning tree is very simple, and does not depend on the ordering chosen on X (Fig. 8.2).

More precisely, we start at x_0 a Markov chain path ω_1, then if n_1 denotes $\inf(i, x_i \notin \omega^{BE}_1)$, we start at x_{n_1} a Markov chain path ω_2 killed as it hits $\{\omega^{BE}_1\}$ or Δ, etc ... until the space is covered by the branches ω^{BE}_j, which form a spanning tree $\tau(\omega)$ rooted in Δ.

Given Υ a spanning tree rooted in Δ, denote by η_1 the geodesic in Υ between x_0 and Δ, then by η_2 the geodesic between $x_{\{\inf\{j,x_j \notin \eta_1\}}$ and $\{\eta_2 \cup \Delta\}$ etc ... Then:

$\mathbb{P}(\tau = \Upsilon) = \mathbb{P}(\omega^{BE}_1 = \eta_1)\mathbb{P}(\omega^{BE}_2 = \eta_2|\omega^{BE}_1 = \eta_1)...$

This law is a probability measure $\mathbb{P}^e_{ST}$ on the set $ST_{X,\Delta}$ of spanning trees of X rooted at the cemetery point Δ defined by the energy e. The weight attached to each oriented link $g = (x,y)$ of $X \times X$ is the conductance and the weight attached to the link (x,Δ) is κ_x which we can also denote by $C_{x,\Delta}$.

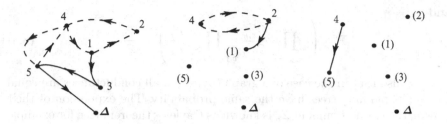

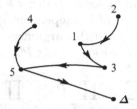

Fig. 8.2 Wilson algorithm

As the determinants simplify in the iteration, the probability of a tree Υ is given by a simple formula:

$$\mathbb{P}^e_{ST}(\Upsilon) = \mathcal{Z}_e \prod_{\xi \in \Upsilon} C_\xi \tag{8.1}$$

It is clearly independent of the ordering chosen initially. Now note that, since we get a probability

$$\mathcal{Z}_e \sum_{\Upsilon \in ST_{X,\Delta}} \prod_{(x,y) \in \Upsilon} C_{x,y} \prod_{x,(x,\Delta) \in \Upsilon} \kappa_x = 1 \tag{8.2}$$

or equivalently

$$\sum_{\Upsilon \in ST_{X,\Delta}} \prod_{(x,y) \in \Upsilon} P^x_y \prod_{x,(x,\Delta) \in \Upsilon} P^x_\Delta = \frac{1}{\prod_{x \in X} \lambda_x \mathcal{Z}_e}$$

Then, it follows that, for any e' for which conductances (including κ') are positive only on links of e,

$$\mathbb{E}^e_{ST}\left(\prod_{(x,y) \in \Upsilon} \frac{P'^x_y}{P^x_y} \prod_{x,(x,\Delta) \in \Upsilon} \frac{P'^x_\Delta}{P^x_\Delta} \right) = \frac{\prod_{x \in X} \lambda_x}{\prod_{x \in X} \lambda'_x} \frac{\mathcal{Z}_e}{\mathcal{Z}_{e'}}$$

and

$$\mathbb{E}^e_{ST}\left(\prod_{(x,y)\in\Upsilon}\frac{C'_{x,y}}{C_{x,y}}\prod_{x,(x,\Delta)\in\Upsilon}\frac{\kappa'_x}{\kappa_x}\right)=\frac{\mathcal{Z}_e}{\mathcal{Z}_{e'}}. \qquad (8.3)$$

Note also that in the case of a graph (i.e. when all conductances are equal to 1), all spanning trees have the same probability. The expression of their cardinal as the determinant $\mathcal{Z}_e$ is known as Cayley's theorem (see for example [29]).

The formula (8.3) shows a kind of duality between random spanning trees and $\mathcal{L}_1$. It can be extended to $\mathcal{L}_k$ for any integer k if we consider the sum (in terms of number of transitions) of k independent spanning trees.

The symmetry is not crucial in this argument. In fact, it shows as well that for any submarkovian transition matrix P',

$$\det(I-P')=\sum_{\Upsilon\in ST_{X,\Delta}}\prod_{(x,y)\in\Upsilon}[P']^x_y\prod_{x,(x,\Delta)\in\Upsilon}(1-\sum_y[P']^x_y)$$

It follows that for any matrix L', with $[L']^x_y\le 0$ for all $x\ne y$ and $[L']^x_x-\sum_y[L']^x_y\ge 0$ for all x, we have:

$$\det(L')=\sum_{\Upsilon\in ST_{X,\Delta}}\prod_{(x,y)\in\Upsilon}[L']^x_y\prod_{x,(x,\Delta)\in\Upsilon}([L']^x_x-\sum_y[L']^x_y)$$

The result extends to any complex matrix L' by the principle of isolated zeros.

In particular, given any 1-form $\omega^{x,y}$, it follows that:

$$\sum_{\Upsilon\in ST_{X,\Delta}}\left(\prod_{(x,y)\in\Upsilon}C_{x,y}e^{i\omega^{x,y}}\prod_{x,(x,\Delta)\in\Upsilon}\kappa_x-\sum_y C_{x,y}(e^{i\omega^{x,y}}-1)\right)=\det(M_\lambda-Ce^{i\omega})$$

or equivalently

$$\mathbb{E}^e_{ST}\left(\prod_{(x,y)\in\Upsilon}e^{i\omega^{x,y}}\prod_{x,(x,\Delta)\in\Upsilon}(1-\frac{1}{\kappa_x}(\sum_y C_{x,y}(e^{i\omega^{x,y}}-1)))\right)=\frac{\mathcal{Z}_e}{\mathcal{Z}_{e,\omega}}.$$

More generally

$$\mathbb{E}^e_{ST}\left(\prod_{(x,y)\in\Upsilon}\frac{C'_{x,y}}{C_{x,y}}e^{i\omega^{x,y}}\prod_{x,(x,\Delta)\in\Upsilon}\frac{1}{\kappa_x}(\kappa'_x-(\sum_y C'_{x,y}(e^{i\omega^{x,y}}-1)))\right)=\frac{\mathcal{Z}_e}{\mathcal{Z}_{e',\omega}}.$$

Exercise 34. Show that more generally, for any tree T rooted in Δ,
$$\mathbb{P}^e_{ST}(\{\Upsilon,\ T \subseteq \Upsilon\}) = \det(G_{|\{T\}-\Delta \times \{T\}-\Delta}) \prod_{\xi \in Edges(T)} C_\xi,\ \{T\} \text{ denoting}$$
the vertex set of T.

(As usual, $C_{x,\Delta} = \kappa_x$. Hint: Run Wilson's algorithm starting from the leaves of T)

Exercise 35. Using Exercise 3, prove Cayley's Theorem: the complete graph K_n has n^{n-2} spanning trees.

The following result follows easily from Proposition 28.

Corollary 6. *The network defined by the random set of loops $\mathcal{L}_W$ constructed in this algorithm is independent of the random spanning tree, and independent of the ordering. It has the same distribution as the network defined by the loops of $\mathcal{L}_1$.*

Remark 21. Note that Proposition 28 and its corollary can be made more precise with the help of Remark 18. The splitting procedure used there with the help of an auxiliary independent set of Poisson Dirichlet variables allows to reconstruct the set of loops $\mathcal{L}_1/\mathcal{L}_1^{\{x\}^c}$ by splitting the first erased loop in the proof of the proposition. Iterating the procedure we can successively reconstruct all sets $\mathcal{L}_1^{\{\eta_m\}^c}/\mathcal{L}_1^{\{\eta_{m+1}\}^c}$ and finally $\mathcal{L}_1/\mathcal{L}_1^{\{\eta\}^c}$. Then, by Wilson algorithm, we can reconstruct $\mathcal{L}_1$.

Let us now consider the *recurrent case.*

A probability is defined on the non oriented spanning trees by the conductances: $\mathbb{P}^e_{ST}((\mathcal{T})$ is defined by the product of the conductances of the edges of $\mathcal{T}$ normalized by the sum of these products on all spanning trees.

Note that any non oriented spanning tree of X along edges of E defines uniquely an oriented spanning tree $I_\Delta(\mathcal{T})$ if we choose a root Δ. The orientation is taken towards the root which can be viewed as a cemetery point. Then, if we consider the associated Markov chain killed as it hits Δ defined by the energy form $e^{\{\Delta\}^c}$, the previous construction yields a probability $\mathbb{P}^{e^{\{\Delta\}^c}}_{ST}$ on spanning trees rooted at Δ which by (8.1) coincides with the image of $\mathbb{P}^e_{ST}$ by I_Δ. This implies in particular that the normalizing factor $\mathcal{Z}_{e_{\{\Delta\}^c}}$ is independent of the choice of Δ as it has to be equal to $(\sum_{T \in ST_X} \prod_{\{x,y\} \in T} C_{x,y})^{-1}$. We denote it by $\mathcal{Z}^0_e$. This factor can also be expressed in terms of the recurrent Green operator G. Recall it is defined as a scalar product on measures of zero mass. The determinant of G is defined as the determinant of its matrix in any orthonormal basis of this hyperplane, with respect to the natural Euclidean scalar product.

Recall that for any $x \neq \Delta$, $G(\varepsilon_x - \varepsilon_\Delta) = -\frac{\langle \lambda, G^{\{\Delta\}^c} \varepsilon_x \rangle}{\lambda(X)} + G^{\{\Delta\}^c} \varepsilon_x$.
Therefore, for any $y \neq \Delta$, $\langle \varepsilon_y - \varepsilon_\Delta, G(\varepsilon_x - \varepsilon_\Delta) \rangle = [G^{\{\Delta\}^c}]^{x,y}$.

The determinant of the matrix $[G^{\{\Delta\}^c}]$, equal to $\mathcal{Z}^0_e$, is therefore also the determinant of G in the basis $\{\delta_x - \delta_\Delta, x \neq \Delta\}$ which is not orthonormal with respect to the natural euclidean scalar product. An easy calculation shows it equals

$$\det\left(\langle\delta_y-\delta_\Delta,\delta_x-\delta_\Delta\rangle_{\mathbb{R}^{|X|}},x,y\neq\Delta\right)\det(G)=|X|\det(G).$$

Exercise 36. Prove that if we set $\alpha_{x_0}(\mathcal{T})=\prod_{(x,y)\in I_{x_0}(\mathcal{T})}P_y^x$, $\sum_{\mathcal{T}\in ST_X}$ $\alpha_{x_0}(\mathcal{T})$ is proportional to λ_{x_0} as x_0 varies in X. More precisely, it equals $K\lambda_{x_0}$, with $K=\frac{\mathcal{Z}_e^0}{\prod_{x\in X}\lambda_x}$. This fact is known as the matrix-tree theorem ([29]).

Exercise 37. Check directly that $\mathcal{Z}_{e_{\{x_0\}^c}}$ is independent of the choice of x_0.

Exercise 38. Given a spanning tree $\mathcal{T}$, we say a subset A is wired iff the restriction of $\mathcal{T}$ to A is a tree.

(a) Let $\tilde{e}_A$ be the recurrent energy form defined on A by the conductances C. Show that $\mathbb{P}_{ST}^e(A\text{ is wired})=\frac{\mathcal{Z}_e^0}{\mathcal{Z}_{e_{A^c}}\mathcal{Z}_{\tilde{e}_A}}$ (Hint: Choose a root in A. Then use Exercise 34 and identity (8.2).

(b) Show that under $\mathbb{P}_{ST}^e$, given that A is wired, the restriction of the spanning tree of X to A and the spanning tree of $A^c\cup\{\Delta\}$ obtained by rooting at an external point Δ the spanning forest induced on A^c by restriction of the spanning tree are independent, with distributions respectively given by $\mathbb{P}_{ST}^{\tilde{e}_A}$ and $\mathbb{P}_{ST}^{e_{A^c}}$.

(c) Conversely, given such a pair, the spanning tree of X can be recovered by attaching to A the roots y_i of the spanning forest of A^c independently, according to the distributions $\frac{C_{y_i,u}}{\sum_{u\in A}C_{y_i,u}}$, $u\in A$.

8.3   The Transfer Current Theorem

Let us come back to the transient case by choosing some root $x_0=\Delta$. As by the strong Markov property, $V_x^y=\mathbb{P}_y(T_x<\infty)V_x^x$, we have $\frac{G^{y,x}}{G^{x,x}}=\frac{V_x^y}{V_x^x}=\mathbb{P}_y(T_x<\infty)$, and therefore

$$\mathbb{P}_{ST}^e((x,y)\in\Upsilon)=\mathbb{P}_x(\gamma_1^{BE}=y)=V_x^x P_y^x\mathbb{P}^y(T_x=\infty)=C_{x,y}G^{x,x}(1-\frac{G^{x,y}}{G^{x,x}}).$$

Directly from the above, we recover Kirchhoff's theorem:

$$\mathbb{P}_{ST}^e(\pm(x,y)\in\Upsilon)=C_{x,y}[G^{x,x}(1-\frac{G^{x,y}}{G^{x,x}})+G^{y,y}(1-\frac{G^{y,x}}{G^{y,y}})]$$

$$=C_{x,y}(G^{x,x}+G^{y,y}-2G^{x,y})=C_{x,y}K^{(x,y),(x,y)}$$

with the notation introduced in Sect. 1.5, and this is clearly independent of the choice of the root.

Exercise 39. Give an alternative proof of Kirchhoff's theorem by using (8.3), taking $C'_{x,y} = sC_{x,y}$ and $C'_{u,v} = C_{u,v}$ for $\{u,v\} \neq \{x,y\}$.

In order to go further, it is helpful to introduce some elements of exterior algebra. Recall that in any vector space E, in terms of the ordinary tensor product $\otimes$, the skew symmetric tensor product $v_1 \wedge v_2 \wedge ... \wedge v_n$ of n vectors $v_1...v_n$ is defined as $\frac{1}{\sqrt{n!}} \sum_{\sigma \in S_n} (-1)^{m(\sigma)} v_{\sigma(1)} \otimes ... \otimes v_{\sigma(n)}$. They generate the n-th skew symmetric tensor power of E, denoted $E^{\wedge n}$. Obviously, $v_{\sigma(1)} \wedge ... \wedge v_{\sigma(n)} = (-1)^{m(\sigma)} v_1 \wedge v_2 \wedge ... \wedge v_n$. If the vector space is equipped with a scalar product $\langle .,. \rangle$, it extends to tensors and $\langle v_1 \wedge v_2 \wedge ... \wedge v_n, v'_1 \wedge v'_2 \wedge ... \wedge v'_n \rangle = \det(\langle v_i, v'_j \rangle)$.

The following result, which generalizes Kirchhoff's theorem, is known as the transfer current theorem (see for example [28, 29]):

Theorem 5. $\mathbb{P}^e_{ST}(\pm \xi_1, ... \pm \xi_k \in \Upsilon) = (\prod_1^k C_{\xi_i}) \det(K^{\xi_i,\xi_j} \; 1 \leq i, j \leq k)$.

Note this determinant does not depend on the orientation of the links.

Proof. Note first that if Υ is a spanning tree rooted in $x_0 = \Delta$ and $\xi_i = (x_{i-1}, x_i)$, $1 \leq i \leq |X| - 1$ are its oriented edges, the measures $\delta_{x_i} - \delta_{x_{i-1}}$ form another basis of the euclidean hyperplane of signed measures with zero charge, which has the same determinant as the basis $\delta_{x_i} - \delta_{x_0}$.

Therefore, $\mathcal{Z}^0_e$ is also the determinant of the matrix of G in this basis, i.e.

$$\mathcal{Z}^0_e = \det(K^{\xi_i,\xi_j} \; 1 \leq i, j \leq |X| - 1)$$

and

$$\mathbb{P}^e_{ST}(\Upsilon) = (\prod_1^{|X|-1} C_{\xi_i}) \det(K^{\xi_i,\xi_j} \; 1 \leq i, j \leq |X| - 1)$$

$$= \det(\sqrt{C_{\xi_i}} K^{\xi_i,\xi_j} \sqrt{C_{\xi_j}} \; 1 \leq i, j \leq |X| - 1).$$

Recall that $\sqrt{C_{\xi_i}} K^{\xi_i,\xi_j} \sqrt{C_{\xi_j}} = \left\langle \alpha^*_{\xi_i} | \Pi | \alpha^*_{\xi_j} \right\rangle_{\mathbb{A}_-}$, where Π denotes the projection on the space of discrete differentials of functions (contained in $\mathbb{A}_-$) and that $\alpha^{*x,y}_{(\eta)} = \pm \frac{1}{\sqrt{C_\eta}}$ if $(x,y) = \pm(\eta)$ and $= 0$ elsewhere.

To finish the proof of the theorem, it is helpful to use the exterior algebra. Note first that for any ONB $e_1, ..., e_{|X|-1}$ of the space of discrete differentials, $\Pi \alpha^*_\xi = \sum \left\langle \alpha^*_\xi | e_j \right\rangle e_j$ and $\mathbb{P}^e_{ST}(\Upsilon) = \det(\langle \alpha^*_{\xi_i} | e_j \rangle)^2 = \left\langle \alpha^*_{\xi_1} \wedge ... \wedge \alpha^*_{\xi_{|X|-1}} | e_1 \wedge ... \wedge e_{|X|-1} \right\rangle^2_{\wedge^{|X|-1} \mathbb{A}_-}$. Therefore

$$\mathbb{P}^e_{ST}(\xi_1,, ..., \xi_k \in \Upsilon)$$

$$= \sum_{\eta_{k+1}, ... \eta_{|X|-1}} \left\langle \alpha^*_{\xi_1} \wedge ... \wedge \alpha^*_{\xi_k} \wedge \alpha^*_{\eta_{k+1}} \wedge ... \wedge \alpha^*_{\eta_{|X|-1}} | e_1 \wedge ... \wedge e_{|X|-1} \right\rangle^2_{\wedge^{|X|-1} \mathbb{A}_-}$$

where the sum is on all edges $\eta_{k+1}, ..., \eta_{|X|-1}$ completing $\xi_1, , ..., \xi_k$ into a spanning tree. It can be extended to all systems of distinct $q = |X| - 1 - k$ edges $\eta' = \{\eta'_1, ..., \eta'_q\}$ as all the additional term vanish. Indeed, an exterior product of $\alpha^*_{\xi_1}$ vanishes as soon as they form a loop. Hence the expression above equals:

$$\sum_{\eta'} \Big(\sum_{i_1 < ... < i_k} \varepsilon_{i_1...i_k} \left\langle \alpha^*_{\xi_1} \wedge ... \wedge \alpha^*_{\xi_k} | e_{i_1} \wedge ... \wedge e_{i_k} \right\rangle \left\langle \alpha^*_{\eta'_1} \wedge ... \wedge \alpha^*_{\eta'_q} | e_{i'_1} \wedge ... \wedge e_{i'_q} \right\rangle \Big)^2$$

where the i'_l are the indices complementing $i_1, ..., i_k$ put in increasing order and $\varepsilon_{i_1...i_k} = (-1)^{(i_1-1)...(i_k-k)}$. Recalling that the α^*_ξ form an orthonormal base of $\mathbb{A}_-$, we see that the sum in η' of each mixed term in the square vanishes and

$$\sum_{\eta'} \left\langle \alpha^*_{\eta'_1} \wedge ... \wedge \alpha^*_{\eta'_q} | e_{i'_1} \wedge ... \wedge e_{i'_q} \right\rangle^2 = 1.$$

Hence we obtain finally:

$$\sum_{i_1 < ... < i_k} \left\langle \alpha^*_{\xi_1} \wedge ... \wedge \alpha^*_{\xi_k} | e_{i_1} \wedge ... \wedge e_{i_k} \right\rangle^2_{\wedge^k \mathbb{A}_-} = \det(\sqrt{C_{\xi_i}} K^{\xi_i, \xi_j} \sqrt{C_{\xi_j}} \; 1 \le i, j \le k).$$

which ends the proof of the theorem. $\qquad\qquad\qquad\qquad\qquad\qquad\qquad\square$

It follows that given any function g on non oriented links,

$$\mathbb{E}^e_{ST}(e^{-\sum_{\xi \in \Upsilon} g(\xi)}) = \mathbb{E}^e_{ST}(\prod_\xi (1 + (e^{-g(\xi)} - 1)1_{\xi \in \Upsilon})$$

$$= 1 + \sum_{k=1}^{|E|} \sum_{\pm\xi_1 \ne \pm\xi_2 \ne ... \ne \pm\xi_k} \prod (e^{-g(\xi_i)} - 1)\mathbb{P}^e_{ST}(\pm\xi_1, ..., \pm\xi_k \in \Upsilon)$$

$$= 1 + \sum_k \sum_{\pm\xi_1 \ne \pm\xi_2 \ne ... \ne \pm\xi_k} \prod (e^{-g(\xi_i)} - 1) \det(K^{\xi_i, \xi_j} \; 1 \le i, j \le k)$$

$$= 1 + \sum Tr((M_{C(e^{-g}-1)}K)^{\wedge k}) = \det(I + KM_{C(e^{-g}-1)})$$

and we have proved the following

Proposition 29. $\mathbb{E}^e_{ST}(e^{-\sum_{\xi \in \Upsilon} g(\xi)}) = \det(I - M_{\sqrt{C(1-e^{-g})}} K M_{\sqrt{C(1-e^{-g})}}).$

Here determinants are taken on matrices indexed by E.

Remark 22. This is an example of the Fermi point processes (also called determinantal point processes) discussed in [49] and [45]. It is determined by the matrix $M_{\sqrt{C}} K M_{\sqrt{C}}$. Note that it follows also easily from the previous proposition that the set of edges which do not belong to the spanning tree also form a Fermi point process defined by the matrix $I - M_{\sqrt{C}} K M_{\sqrt{C}}$.

In particular, under $\mathbb{P}_{ST}$, the set of points x such that $(x, \Delta) \in \Upsilon$ (i.e. the set of points directly connected to the root Δ) is a Fermi point process the law of which is determined by the matrix $Q^{x,y} = \sqrt{\kappa_x} G^{x,y} \sqrt{\kappa_y}$.

For example, if X is an interval of $\mathbb{Z}$, with $C_{x,y} = 0$ iff $|x - y| > 1$, it is easily verified that for $x < y < z$,

$$Q^{x,z} = \frac{Q^{x,y} Q^{y,z}}{Q^{y,y}}$$

Then using the remark following Theorem 6 in [49], we see that the spacings of this point process are independent (in the sense that given any point of the process, the spacings on his right and on his left are independent).

The edges of X which do not belong to the spanning tree form a determinantal process of edges, of the same type, intertwinned with the points connected to Δ.

A consequence is that for any spanning tree T, if π_T denotes $M_{1_{\{T\}}}$ (the multiplication by the indicator function of T), it follows from the above, by letting g be $m 1_{\{T^c\}}$, $m \to \infty$ that

$$\mathbb{P}^e_{ST}(T) = \det((I - KM_C)(I - \pi_T) + \pi_T) = \det((I - KM_C)_{T^c \times T^c}).$$

Another consequence is that if e' is another energy form on the same graph,

$$\mathbb{E}^e_{ST}\left(\prod_{(x,y) \in \Upsilon} \frac{C'_{x,y}}{C_{x,y}} \right) = \det(I - M_{\sqrt{C-C'}} KM_{\sqrt{C-C'}}).$$

On the other hand, from (8.3), it also equals

$$\left(\sum_{T \in ST_X} \prod_{\{x,y\} \in T} C_{x,y} \right) \mathcal{Z}^0_e = \frac{\mathcal{Z}^0_e}{\mathcal{Z}^0_{e'}}$$

so that finally

$$\frac{\mathcal{Z}^0_e}{\mathcal{Z}^0_{e'}} = \det(I - M_{\sqrt{C-C'}} KM_{\sqrt{C-C'}}).$$

Note that indicators of distinct individual edges are negatively correlated. More generally:

Theorem 6. *(Negative association) Given any sets disjoint of edges E_1 and E_2,*

$$\mathbb{P}^e_{ST}(E_1 \cup E_2 \subseteq \Upsilon) \leq \mathbb{P}^e_{ST}(E_1 \subseteq \Upsilon)\mathbb{P}^e_{ST}(E_2 \subseteq \Upsilon).$$

Proof. Denote by $K^{\#}(i,j)$ the restriction of $K^{\#} = (\sqrt{C_\xi} K^{\xi,\eta} \sqrt{C_\eta},\ \xi, \eta \in E)$ to $E_i \times E_j$. Then,

$$\frac{\mathbb{P}^e_{ST}(E_1 \cup E_2 \subseteq \Upsilon)}{\mathbb{P}^e_{ST}(E_1 \subseteq \Upsilon)\mathbb{P}^e_{ST}(E_2 \subseteq \Upsilon)} = \frac{\det(K^{\#})}{\det(K^{\#}(2,2))\det(K^{\#}(2,2))} = \det\left(\begin{bmatrix} I & F \\ F^* & I \end{bmatrix}\right)$$

with $F = K^{\#}(1,1)^{-\frac{1}{2}} K^{\#}(1,2) K^{\#}(2,2)^{-\frac{1}{2}}$

Finally, note that $\log(\det\left(\begin{bmatrix} I & F \\ F^* & I \end{bmatrix}\right)) = Tr(\log\left(\begin{bmatrix} I & F \\ F^* & I \end{bmatrix}\right))$

$= -\sum_1^\infty \frac{1}{2k} Tr((FF^*)^k) \leq 0.$ $\qquad\square$

Remark 23. Note that it follows directly from the expression of P_{ST} and from the transfer current theorem that for any set of disjoint edges $\xi_1, ..., \xi_k$:

$$[\mathcal{Z}^0_e]^{-1} \frac{\partial^k}{\partial C_{\xi_1}...\partial C_{\xi_k}} [\mathcal{Z}^0_e]^{-1} = \det(K_{\xi_i,\xi_j}, 1 \leq i, j \leq k).$$

Proof. Note that

$$\frac{\partial k}{\partial C_{\xi_1}...\partial C_{\xi_k}} [\mathcal{Z}^0_e]^{-1} = \frac{\partial^k}{\partial C_{\xi_1}...\partial C_{\xi_k}} \sum_{T \in ST_X} \prod_{\{x,y\} \in T} C_{x,y}$$

$$= [\mathcal{Z}^0_e \prod_1^k C_{\xi_i}]^{-1} \mathbb{P}^e_{ST}(\pm\xi_1, ..., \pm\xi_k \in \Upsilon).$$

$\qquad\square$

This result can be proved directly using for example Grassmann variables (as used in [23]). The transfer current theorem can then be derived immediately from it as shown in the following section.

8.4   The Skew-Symmetric Fock Space

Consider the real Fermionic Fock space $\Gamma^\wedge(\mathbb{H}^*) = \overline{\oplus\mathbb{H}^{*\wedge n}}$ obtained as the closure of the sum of all skew-symmetric tensor powers of $\mathbb{H}^*$ (the zero-th tensor power is $\mathbb{R}$). In terms of the ordinary tensor product $\otimes$, the skew-symmetric tensor product $\mu_1 \wedge ... \wedge \mu_n$ is defined as $\frac{1}{\sqrt{n!}} \sum_{\sigma \in \mathcal{S}_n} (-1)^{m(\sigma)} \mu_{\sigma(1)} \otimes ... \otimes \mu_{\sigma(n)}$. The construction of $\Gamma(\mathbb{H}^*)$ is known as Fermi second quantization.

For any $x \in X$, the annihilation operator c_x and the creation operator c_x^* are defined as follows, on the uncompleted Fock space $\oplus\mathbb{H}^{*\wedge n}$:

$$c_x(\mu_1 \wedge \ldots \wedge \mu_n) = (-1)^{k-1} \sum_k G\mu_k(x)\mu_1 \wedge \ldots \wedge \mu_{k-1} \wedge \mu_{k+1} \wedge \ldots \wedge \mu_n$$

$$c_x^*(\mu_1 \wedge \ldots \wedge \mu_n) = \delta_x \wedge \mu_1 \wedge \ldots \wedge \mu_n$$

Note that c_y^* is the dual of c_y and that $[c_x, c_y^*]^+ = G(x, y)$ with all others anticommutators vanishing.

We will work on the complex Fermionic Fock space $\mathcal{F}_F$ defined as the tensor product of two copies of $\Gamma^\wedge(\mathbb{H}^*)$ which is isomorphic to $\Gamma^\wedge(\mathbb{H}_1^* + \mathbb{H}_2^*)$, $\mathbb{H}_1$ and $\mathbb{H}_2$ being two copies of $\mathbb{H}$. The complex Fock space structure is defined by two anticommuting sets of creation and annihilation operators. $\mathcal{F}_F$ is generated by the vector 1 and creation/annihilation operators c_x, c_x^*, d_x, d_x^* with $[c_x, c_y^*]^+ = [d_x, d_y^*]^+ = G(x, y)$ and with all others anticommutators vanishing. Here, c_x^* creates δ_x in $\mathbb{H}_1^*$ and d_x^* creates δ_x in $\mathbb{H}_2^*$.

Anticommuting variables $\psi^x, \overline{\psi}^x$ are defined as operators on the Fermionic Fock space $\mathcal{F}_F$ by:

$$\psi^x = \sqrt{2}(d_x + c_x^*) \text{ and } \overline{\psi}^x = \sqrt{2}(-c_x + d_x^*).$$

The following anticommutation relations hold

$$[\psi^x, \psi^y]^+ = [\overline{\psi}^x, \psi^y]^+ = [\overline{\psi}^x, \overline{\psi}^y]^+ = 0$$

In particular, $(\sum_x \lambda_x \psi^x)^2 = (\sum_x \lambda_x \overline{\psi}^x)^2 = 0$. Note that $\overline{\psi}_x$ is not the dual of ψ_x, but there is an involution $\mathfrak{I}$ on $\mathcal{F}_F$ such that $\overline{\psi} = \mathfrak{I}\psi^*\mathfrak{I}$.

$\mathfrak{I}$ is defined by its action on each tensor power: it multiplies each element in $\mathbb{H}^{*\wedge m} \otimes \mathbb{H}^{*\wedge p}$ by $(-1)^m$.

Exercise 40. Show that in contrast with the Bosonic case, all these operators are bounded.

Simple calculations yield that:

$$\left\langle 1, \psi^{x_m} \ldots \psi^{x_1} \overline{\psi}^{y_1} \ldots \overline{\psi}^{y_n} 1 \right\rangle = \delta_{nm} 2^n \det(G(x_i, y_j)).$$

Therefore

$$\left\langle 1, \psi^{x_n} \overline{\psi}^{y_n} \ldots \psi^{x_1} \overline{\psi}^{y_1} 1 \right\rangle = 2^n \det(G(x_i, y_j)).$$

and

$$\left\langle 1, \exp(-\frac{1}{2}e(\psi, \overline{\psi}) + \frac{1}{2}e'(\psi, \overline{\psi}))1 \right\rangle_{\mathcal{F}_F} = \frac{\det(G)}{\det(G')}.$$

Indeed, if f_i is an orthonormal basis of $\mathbb{H}$, in which e' is diagonal with eigenvalues λ_i, the first side equals $\left\langle 1, \prod_i \exp(\frac{1}{2}(-1 + \lambda_i) \langle \psi, f_i \rangle \langle \overline{\psi}, f_i \rangle)1 \right\rangle_{\mathcal{F}_F} = \left\langle 1, \prod_i (1 + \frac{1}{2}(-1 + \lambda_i) \langle \psi, f_i \rangle \langle \overline{\psi}, f_i \rangle)1 \right\rangle_{\mathcal{F}_F} = 1 + \sum_k \sum_{i_1 < \ldots i_k} (-1 + \lambda_{i_1}) \ldots (-1 + \lambda_i) = \prod \lambda_i$.

In particular, for any positive measure χ on X,

$$\left\langle 1, \exp(\sum_x \chi_x \psi^x \overline{\psi}^x) 1 \right\rangle_{\mathcal{F}_F} = \frac{\det(G)}{\det(G_\chi)} = \left\langle 1, \exp(-\sum_x \chi_x \varphi^x \overline{\varphi}^x) 1 \right\rangle_{\mathcal{F}_B}^{-1}.$$

We observe a "Supersymmetry" between ϕ and ψ: for any exponential or polynomial F

$$\left\langle 1, F(\phi\overline{\phi} - \psi\overline{\psi}) 1 \right\rangle_{\mathcal{F}_B \otimes \mathcal{F}_F} = F(0).$$

(1 denotes $1_{(B)} \otimes 1_{(F)}$)

Remark 24. On a finite graph, $\psi, \overline{\psi}$ and the whole supersymmetric complex Fock space structure can also be defined in terms of complex differential forms defined on $\mathbb{C}^{|X|}$, using exterior products, interior products and De Rham $*$ operator. This extension of the Gaussian representation of the complex Bosonic Fock space is explained in the introduction of [23]. It was used for example in [26].

Note that

$$E_{ST}\left(\prod_{(x,y) \in \tau} \frac{C'_{x,y}}{C_{x,y}} \prod_{x,(x,\delta) \in \tau} \frac{\kappa'_x}{\kappa_x} \right) = \frac{\mathcal{Z}_e}{\mathcal{Z}_{e'}} = \left\langle 1, \exp(-\frac{1}{2}e(\psi, \overline{\psi}) + \frac{1}{2}e'(\psi, \overline{\psi})) 1 \right\rangle_{\mathcal{F}_F}.$$

The Transfer Current Theorem follows directly, by calculation of

$$P_{ST}((x_i, y_i) \in \tau) = \prod C_{x_i, y_i} \frac{\partial^k}{\partial C'_{x_1, y_1} \cdots \partial C'_{x_k, y_k}} \Big|_{C'=C}$$

$$\times \left\langle 1, \exp(-\frac{1}{2}e(\psi, \overline{\psi}) + \frac{1}{2}e'(\psi, \overline{\psi})) 1 \right\rangle_{\mathcal{F}_F})$$

$$= 2^{-k} \prod C_{x_i, y_i} \left\langle 1, (\prod (\psi^{y_i} - \psi^{x_i})(\overline{\psi}^{y_i} - \overline{\psi}^{x_i}) 1 \right\rangle_{\mathcal{F}_F})$$

$$= \det(K^{(x_i, y_i), (x_j, y_j)}) \prod C_{x_i, y_i}.$$

Note that in the second line we use the identity $C_{x,y} = C_{y,x}$ to show that $\frac{\partial}{\partial C'_{x,y}} e'(\psi, \overline{\psi}) = (\psi^y - \psi^x)(\overline{\psi}^y - \overline{\psi}^x)$.

In particular, $P_{ST}((x_i, \delta) \in \tau) = \det(G(x_i, x_j)) \prod \kappa_{x_i}$.

More generally, we can show that

$$\frac{\mathcal{Z}_e}{\mathcal{Z}_{e',\omega}} = E_{ST}\left(\prod_{(x,y) \in \tau} \frac{C'_{x,y}}{C_{x,y}} e^{i\omega_{x,y}} \prod_{x,(x,\delta) \in \tau} \frac{\kappa'_x - (\sum_y C'_{x,y}(e^{i\omega^{x,y}} - 1))}{\kappa_x} \right)$$

and

$$\frac{\mathcal{Z}_e}{\mathcal{Z}_{e',\omega}} = \left\langle 1, \exp(\frac{1}{2}e(\psi,\overline{\psi}) - \frac{1}{2}e'^{(\omega)}(\psi,\overline{\psi}))1 \right\rangle_{\mathcal{F}_F}.$$

Hence, if (x_j, y_j) are disjoint, and Υ denotes the random spanning tree rooted in Δ, it follows that

$$\prod(\frac{\partial}{\partial C'_{x_j,y_j}} - \frac{i}{C_{x_j,y_j}}\frac{\partial}{\partial \omega_{x_j,y_j}}) + \frac{\partial}{\partial \kappa'_{x_j}})|_{C'=C,\omega=0,\kappa'=\kappa}\frac{\mathcal{Z}_e}{\mathcal{Z}_{e',\omega}} = \frac{2}{C_{x_j,y_j}}P_{ST}((x_j,y_j) \in \Upsilon).$$

Therefore, we get the analogue of the transfer current theorem for rooted spanning trees:

$$P_{ST}((x_j,y_j) \in \Upsilon) = \prod 2^{-k}C_{x_j,y_j}\left\langle 1, (\prod \psi^{x_j}(\overline{\psi}^{y_j} - \overline{\psi}^{x_j})1 \right\rangle_{\mathcal{F}_F})$$

$$= \det(G^{x_j,y_k} - G^{x_j,x_k})\prod C_{x_j,y_j}$$

The relations we have established can be summarized in the following diagram:

(Wilson Algorithm)

Loop ensemble $\mathcal{L}_1$ $\longleftrightarrow$ Random Spanning Tree

$\updownarrow$ $\updownarrow$

Free field $\phi, \overline{\phi}$ $\longleftrightarrow$ Grassmann field $\psi, \overline{\psi}$

("Supersymmetry")

NB: ϕ and ψ can also be used jointly to represent bridge functionals (Cf. [23]): in particular

$$\int F(\tilde{l})\mu_{x,y}(dl) = \left\langle 1, \phi_x\overline{\phi}_y F(\phi\overline{\phi} - \psi\overline{\psi})1 \right\rangle_{\mathcal{F}_B \otimes \mathcal{F}_F} = \left\langle 1, \psi_x\overline{\psi}_y F(\phi\overline{\phi} - \psi\overline{\psi})1 \right\rangle_{\mathcal{F}_B \otimes \mathcal{F}_F}.$$

Chapter 9
Reflection Positivity

9.1 Main Result

In this section, we assume there exists a partition of X: $X = X^+ \cup X^-$, $X^+ \cap X^- = \varnothing$ and an involution ρ on X such that:

(a) e is ρ-invariant.
(b) ρ exchanges X^+ and X^-.
(c) The $X^+ \times X^+$ matrix $C_{x,y}^{\pm} = C_{x,\rho(y)}$, is nonnegative definite.

Then the following holds:

Theorem 7. *(i) For any positive integer d and square integrable function Φ in $\sigma(\widehat{\mathcal{L}_d}^x, x \in X^+) \vee \sigma(N_{x,y}^{(d)}, x, y \in X^+)$,*

$$\mathbb{E}(\Phi(\mathcal{L}_d)\overline{\Phi}(\rho(\mathcal{L}_d))) \geq 0.$$

(ii) For any square integrable function Σ of the free field ϕ restricted to X^+,

$$\mathbb{E}_\phi(\Sigma(\phi)\overline{\Sigma}(\rho(\phi))) \geq 0.$$

(iii) For any set of edges $\{\xi_i\}$ in $X^+ \times X^+$ the matrix,

$$K_{i,j} = \mathbb{P}_{ST}(\xi_i \in T, \rho\xi_j \in T) - \mathbb{P}_{ST}(\xi_i \in T)\mathbb{P}_{ST}(\xi_j \in T)$$

is nonpositive definite.

Proof. The property (ii) is well known in a slightly different context and is named reflexion positivity: Cf. for example [11, 48] and their references. Reflection positivity is a keystone in the bridge between statistical and quantum mechanics.

To prove (i), we use the fact that the σ-algebra is generated by the algebra of random variables of the form $\Phi = \sum \lambda_j B_{(d)}^{e,e_j,\omega_j}$ with $C^{(e_j)} = C$ and $\omega_j = 0$

Y. Le Jan, *Markov Paths, Loops and Fields*, Lecture Notes in Mathematics 2026, DOI 10.1007/978-3-642-21216-1_9, © Springer-Verlag Berlin Heidelberg 2011

except on $X^+ \times X^+$, $C^{(e_j)} \leq C$ on $X^+ \times X^+$, $\lambda^{(e_j)} = \lambda$ on X^- and $\lambda^{(e_j)} \geq \lambda$ on X^+.
Then

$$\mathbb{E}(\Phi(\mathcal{L}_d)\overline{\Phi}(\rho(\mathcal{L}_d))) = \mathbb{E}(\sum \lambda_j \overline{\lambda}_q B_{(d)}^{e,e_{j,q},\omega_j - \rho(\omega_q)}) = \sum \lambda_j \overline{\lambda}_q (\frac{\mathcal{Z}_{e_{j,q},\omega_j - \rho(\omega_q)}}{\mathcal{Z}_e})^d$$

with $e_{j,q} = e_j + \rho(e_q) - e$.

We have to prove this is non negative. It is enough to prove it for $d = 1$, as the Hadamard product of two nonnegative definite Hermitian matrices is nonnegative definite.

Let us first assume that the nonnegative definite matrix $C^{\pm}$ is positive definite. We will see that the general case can be reduced to this one.

Now note that $\mathcal{Z}_{e_j + \rho(e_q) - e, \omega_j - \rho(\omega_q)}$ is the inverse of the determinant of a positive definite matrix of the form:

$$D(j,q) = \begin{bmatrix} A(j) & -C^{\pm} \\ -C^{\pm} & A(q)^* \end{bmatrix}$$

with $[A(j)]_{u,v} = \lambda_u^{(e_j)} \delta_{u,v} - C_{u,v}^{(e_j)} e^{i\omega_j^{u,v}}$ and $C_{u,v}^{\pm} = C_{u,\rho(v)}$.
It is enough to show that $\det(D(j,k))^{-1}$ can be expanded in series of products $\sum q_n(j)\overline{q}_n(k)$ with $\sum |q_n(j)|^2 < \infty$.
As

$$D(j,q) =$$

$$\begin{bmatrix} [C^{\pm}]^{\frac{1}{2}} & 0 \\ 0 & [C^{\pm}]^{\frac{1}{2}} \end{bmatrix} \begin{bmatrix} [C^{\pm}]^{-\frac{1}{2}} A(j)[C^{\pm}]^{-\frac{1}{2}} & -I \\ -I & [C^{\pm}]^{-\frac{1}{2}} A(q)^*[C^{\pm}]^{-\frac{1}{2}} \end{bmatrix} \begin{bmatrix} [C^{\pm}]^{\frac{1}{2}} & 0 \\ 0 & [C^{\pm}]^{\frac{1}{2}} \end{bmatrix}$$

the inverse of this determinant can be written

$$\det(C^{\pm})^{-2} \det(F(j)) \det(F(q)^*) \det(I - \begin{bmatrix} 0 & F(j) \\ F(q)^* & 0 \end{bmatrix})^{-1}$$

with $F(j) = [C^{\pm}]^{\frac{1}{2}} A(j)^{-1}[C^{\pm}]^{\frac{1}{2}}$, or more simply:

$$F(j) = \det(A(j))^{-1} \det(A(q)^*)^{-1} \det(I - \begin{bmatrix} 0 & F(j) \\ F(q)^* & 0 \end{bmatrix})^{-1}.$$

Note that $A(j)^{-1}$ is also the Green function of the restriction to X^+ of the Markov chain associated with e_j, twisted by ω_j. Therefore $A(j)^{-1}C^{\pm} = [C^{\pm}]^{-\frac{1}{2}} F(j)[C^{\pm}]^{\frac{1}{2}}$ is the balayage kernel on X^- defined by this Markov chain with an additional phase under the expectation produced by ω_j. It is therefore clear that the eigenvalues of the matrices $A(j)^{-1}C^{\pm}$ and $F(j)$ are of modulus less than one and it follows that

$$\begin{bmatrix} 0 & F(j) \\ F(q)^* & 0 \end{bmatrix} = \begin{bmatrix} 0 & I \\ I & 0 \end{bmatrix} \begin{bmatrix} F(q)^* & 0 \\ 0 & F(j) \end{bmatrix}$$

is a contraction. We can always assume it is a strict contraction, by adding a killing term we can let converge to zero once the inequality is proved.

If X^+ has only one point, $(1 - F(j)F(q)^*)^{-1} = \sum F(j)^{-n}\overline{F(q)}^{-n}$ which allows to conclude. Let us now treat the general case.

For any (n, m) matrix N, and $k = (k_1, ..., k_m) \in \mathbb{N}^m$, $l = (l_1, ..., l_n) \in \mathbb{N}^n$, let $N^{\{k,l\}}$ denote the $(|k|, |l|)$ matrix obtained from by repeating k_i times each line i; then l_j times each column j.

We use the expansion

$$\det(I - M)^{-1} = 1 + \sum \frac{1}{|k|!} Per(M^{\{k,k\}})$$

valid for any strict contraction M (Cf. [54] and [55]).

Note that if X has $2d$ points, if we denote $(k_1, ..., k_{2d})$ by (k^+, k^-), with $k^+ = (k_1, ..., k_d)$ and $k^- = (k_{d+1}, ..., k_{2d})$,

$$\begin{bmatrix} 0 & F(j) \\ F(q)^* & 0 \end{bmatrix}^{\{k,k\}} = \begin{bmatrix} 0 & F(j)^{\{k^+,k^-\}} \\ [F(q)^*]^{\{k^-,k^+\}} & 0 \end{bmatrix}.$$

But the all terms in the permanent of a $(2n, 2n)$ matrix of the form $\begin{bmatrix} 0 & A \\ B^* & 0 \end{bmatrix}$ vanish unless the submatrices A and B are square matrices (not necessarily of equal ranks). Hence in our case, we necessary have $|k^+| = |k^-|$, so that, A and B are (n, n) matrices.

Then, the non zero terms in the permanent come from permutations exchanging $\{1, 2, ..., n\}$ and $\{n + 1, ..., 2n\}$, which can be decomposed into a pair of permutations of $\{1, 2, ..., n\}$. Therefore:

$$Per(\begin{bmatrix} 0 & A \\ B^* & 0 \end{bmatrix}) = Per(A)Per(B^*)$$

which concludes the proof in the positive definite case as

$$Per(B^*) = \sum_{\tau \in \mathcal{S}_n} \prod_1^n B_{i,\tau(i)}^* = \sum_{\tau \in \mathcal{S}_n} \prod_1^n \overline{B_{\tau(i),i}} = \overline{Per(B)}.$$

To treat the general case where $C^{\pm}$ is only nonnegative definite., we can use a passage to the limit or alternatively, the Proposition 26 (or more precisely its extension including a current) to reduce the sets X^+ and X^- to the support of $C^\pm$.

To prove (ii) let us first show the assumptions imply that the $X^+ \times X^+$ matrix $G_{x,y}^\pm = G^{x,\rho(y)}$ is also nonnegative definite. Let us write G in the form

$$\begin{bmatrix} A & -C^\pm \\ -C^\pm & A \end{bmatrix}^{-1} \quad \text{with } A = M_\lambda - C. \text{ Then}$$

$$G = \begin{bmatrix} A^{-\frac{1}{2}} & 0 \\ 0 & A^{-\frac{1}{2}} \end{bmatrix} \begin{bmatrix} I & -A^{-\frac{1}{2}}C^\pm A^{-\frac{1}{2}} \\ -A^{-\frac{1}{2}}C^\pm A^{-\frac{1}{2}} & I \end{bmatrix}^{-1} \begin{bmatrix} A^{-\frac{1}{2}} & 0 \\ 0 & A^{-\frac{1}{2}} \end{bmatrix}.$$

$A^{-\frac{1}{2}}C^\pm A^{-\frac{1}{2}}$ is non negative definite and as before, we can check it is a contraction since $A^{-1}C^\pm$ is a balayage kernel.

Note that if a symmetric nonnegative definite matrix K has eigenvalues μ_i, the eigenvalues of the symmetric matrix E defined by

$$\begin{bmatrix} I & -K \\ -K & I \end{bmatrix}^{-1} = \begin{bmatrix} D & E \\ E & D \end{bmatrix}$$

are easily seen (exercise) to be $\frac{\mu_i}{1-\mu_i^2}$. Taking $K = A^{-\frac{1}{2}}C^\pm A^{-\frac{1}{2}}$, it follows that the symmetric matrix E, (and in our particular case $G^\pm = A^{-\frac{1}{2}}EA^{-\frac{1}{2}}$) is nonnegative definite.

To finish the proof, let us take Σ of the form $\sum \lambda_j e^{\langle \phi, \chi_j \rangle}$. Then

$$\mathbb{E}_\phi(\Sigma(\phi)\overline{\Sigma}(\rho(\phi))) = \sum \lambda_j \lambda_q \mathbb{E}_\phi(e^{\langle \phi,\chi_j \rangle + \langle \phi, \rho(\chi_q) \rangle})$$

$$= \sum \lambda_j e^{\frac{1}{2}\langle \chi_j, G^{++}\chi_j \rangle} \lambda_k e^{\frac{1}{2}\langle \chi_q, G^{++}\chi_q \rangle} e^{\langle \chi_j, G^\pm \chi_q \rangle}$$

(using that $G^\pm$ is symmetric).

As $G^\pm$ is positive definite, we can conclude since $e^{\frac{1}{2}\langle \chi_j, G^\pm \chi_q \rangle} = \mathbb{E}_w(e^{\langle w, \chi_j \rangle} e^{\langle w, \chi_q \rangle})$, w denoting the Gaussian field on X^+ with covariance $G^\pm$.

To prove (iii), note that the transfer impedance matrix can be decomposed as G. In particular, $K_{i,j} = -(K_{\xi_i,\xi_j}^\pm)^2$, with

$$K_{(x,y),(u,v)}^\pm = K^{(x,y),(\rho(u),\rho(v))} = G^\pm(x,u) + G^\pm(y,v) - G^\pm(x,v) - G^\pm(y,u).$$

Then, using again the Gaussian vector w, and the Wick squares of its components:

$$(K_{(x,y),(u,v)}^\pm)^2 = E(:(w_u - w_v)^2 :: (w_x - w_y)^2 :).$$

$\square$

Remark 25. (a) If U_j are unitary representations with $d_U = d$ and such that $U_j^{x,y}$ is the identity outside $X^+ \times X^+$, (i) can be extended to variables of the form $\sum \lambda_j B_{(d)}^{e,e_j,U_j}$ and to the σ-field they generate.

(b) The property (i) can be also derived from the reflection positivity of the free field (ii) and by Remark 13. Then it can also be proved that for any set of points $\{x_i\}$ in X^+, the matrix $\mathbb{E}(\Phi(\widehat{\mathcal{L}_d})\overline{\Phi}(\rho(\widehat{\mathcal{L}_d})N_{x_i,\rho x_j})$ is non-negative definite.

(c) In the case where α is a half integer, by Remark 11, the reflection positivity of the free field (ii), implies (i) holds also for any *half integer* α provided that $\Phi \in \sigma(\widehat{\mathcal{L}_\alpha}^x, x \in X^+) \vee \sigma(N_{x,y}^{(\alpha)} + N_{y,x}^{(\alpha)}, x, y \in X^+)$.

Exercise 41. Prove the above remarks.

Remark 26. If there exists a partition of X: $X = X^+ \cup X^- \cup X^0$, and an involution ρ on X such that:

(a) e and X^0 are ρ-invariant
(b) $\rho(X^\pm) = X^\mp$
(c) X^+ and X^- are disconnected

Then the assumptions of the previous theorem are satisfied for the trace on $X^+ \cup X^-$.

Moreover, if $X^0 \times X^0$ does not contain any edge of the graph, the assertion (i) of Theorem 7 holds for the non disjoint sets $X^+ \cup X^0$ and $X^- \cup X^0$. More precisely, (i), holds for Φ in $\sigma(\widehat{\mathcal{L}_d}^x, x \in X^+ \cup X^0) \vee \sigma(N_{x,y}^{(d)}, x, y \in X^+ \cup X^0)$. It is enough to apply the theorem to the graph obtained by duplication of each point x_0 in X^0 into (x_0^+, x_0^-), with $x_0^\pm$ connected to points in $X^\pm$ and connected together by conductances $C_{x_0^+, x_0^-}$ we can let increase to infinity.

9.2 A Counter Example

Let show that the reflexion positivity does not hold under μ for loop functionals. Therefore, it will be clear it does not hold for small α. We will consider functionals of the occupation field.

Consider the graph formed by a cube $\pm a$, $\pm b$, $\pm c$, $\pm d$ and the mid-points $\pm \alpha$, $\pm \beta$, $\pm \gamma$, $\pm \delta$ of the sides $\pm ab$, $\pm cd$, $\pm ac$, $\pm bd$. The edges are given by the sides of the cube, as in the picture below (Fig. 9.1).

We can take for example all conductances and killing rates to be equal. Then the symmetry $\rho : x \to -x$ defines an involution satisfying the assumption of Theorem 7. Define the set of loops $A = \{l, \widehat{l^\alpha}\widehat{l^\beta} > 0\}$, $A' = \{l, \widehat{l^\alpha} = \widehat{l^\beta} = 0\}$, $B = \{l, \widehat{l^\gamma}\widehat{l^\delta} > 0\}$ and $B' = \{l, \widehat{l^\gamma} = \widehat{l^\delta} = 0\}$. Note that $A \cap B' \cap \rho(A) \cap \rho(B')$, $A' \cap B \cap \rho(A') \cap \rho(B)$ are empty. But

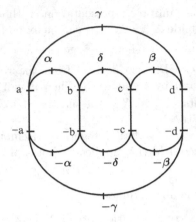

Fig. 9.1 A counter example

$A' \cap B \cap \rho(A) \cap \rho(B')$ and $A \cap B' \cap \rho(A') \cap \rho(B)$ are not (consider the loop $a\alpha b(-b)(-\delta)(-c)c\beta d(-d)(-\gamma)(-a)a$).

Then, if we set $\Phi = 1_{A \cap B'} - 1_{A' \cap B}$, it is clear that

$$\mu(\Phi.\Phi \circ \rho) = -2\mu(A \cap B' \cap \rho(A') \cap \rho(B)) < 0.$$

9.3 Physical Hilbert Space and Time Shift

We will now work under the assumptions of Remark 26, namely, without assuming that $X = X^+ \cup X^-$.

The following results and terminology are inspired by methods of constructive Quantum field theory (Cf. [48] and [11]).

Let $\mathcal{H}^+$ be the space of square integrable functions in $\sigma((\widehat{\mathcal{L}_1}^x, x \in X^+) \vee \sigma(N_{x,y}^{(1)}, x, y \in X^+)$, equipped with the scalar product $\langle \Phi, \Psi \rangle_{\mathcal{H}} = \mathbb{E}(\Phi(\mathcal{L}_1)\Psi(\rho(\mathcal{L}_1)))$ Note that $\langle \Phi, \Phi \rangle_{\mathcal{H}} \leq \mathbb{E}(\Phi^2(\mathcal{L}_1))$ by Cauchy–Schwartz inequality.

Let $\mathcal{N}$ be the subspace $\{\Psi \in \mathcal{H}^+, \mathbb{E}(\Psi(\mathcal{L}_1)\Psi(\rho(\mathcal{L}_1))) = 0\}$ and $\mathcal{H}$ the closure (for the topology induced by this scalar product) of the quotient space $\mathcal{H}^+/\mathcal{N}$ (which can be called the physical Hilbert space). We denote $\Phi^\sim$ the equivalence class of Φ. $\mathcal{H}$ is equipped with the scalar product defined unambiguously by $\langle \Phi^\sim, \Psi^\sim \rangle_{\mathcal{H}} = \langle \Phi, \Psi \rangle_{\mathcal{H}}$.

Assume X is of the form $X_0 \times \mathbb{Z}$ (space $\times$ time)) and let θ be the natural time shift. We assume θ preserves e, i.e. that conductances and κ are θ-invariant. We define ρ by $\rho(x_0, n) = (x_0, -n)$ and assume e is ρ-invariant. Note that $\theta(X^+) \subseteq X^+$ and $\rho\theta = \theta^{-1}\rho$. The transformations

ρ and θ induce a transformations on loops that preserves μ, and θ induces a linear transformation of $\mathcal{H}^+$. Moreover, given any F in $\mathcal{N}$, $F \circ \theta \in \mathcal{N}$, as $\langle F \circ \theta, F \circ \theta \rangle_{\mathcal{H}}$ is nonnegative and equals

$$\mathbb{E}(F \circ \theta(\mathcal{L}_1)F \circ \theta(\rho(\mathcal{L}_1)) = \mathbb{E}(F(\theta(\mathcal{L}_1))F(\rho \circ \theta^{-1}(\mathcal{L}_1)) = \mathbb{E}(F(\theta^2(\mathcal{L}_1))F(\rho(\mathcal{L}_1))$$

$$= \langle F \circ \theta^2, F \rangle_{\mathcal{H}} \le \sqrt{\langle F \circ \theta^2, F \circ \theta^2 \rangle_{\mathcal{H}} \langle F, F \rangle_{\mathcal{H}}}$$

which vanishes.

Proposition 30. *There exist a self adjoint contraction of $\mathcal{H}$, we will denote $\Pi^{(\theta)}$ such that $[\Phi \circ \theta]^{\sim} = \Pi^{(\theta)}(\Phi^{\sim})$.*

Proof. The existence of $\Pi^{(\theta)}$ follows from the last observation made above. As θ preserves μ, it follows from the identity $\rho\theta = \theta^{-1}\rho$ that

$$\langle F \circ \theta, G \rangle_{\mathcal{H}} = \mathbb{E}(F(\theta(\mathcal{L}_1))G(\rho(\mathcal{L}_1)) = \mathbb{E}(F(\mathcal{L}_1)G(\rho \circ \theta^{-1}(\mathcal{L}_1)) = \mathbb{E}(F(\mathcal{L}_1)G(\theta \circ \rho(\mathcal{L}_1))$$

$$= \mathbb{E}(F(\rho(\mathcal{L}_1))G(\theta(\mathcal{L}_1)) = \langle G \circ \theta, F \rangle_{\mathcal{H}}.$$

Therefore, $\Pi^{(\theta)}$ is self adjoint on $\mathcal{H}^+/\mathcal{N}$. To prove that it is a contraction, it is enough to show that $\langle F \circ \theta, F \circ \theta \rangle_{\mathcal{H}} \le \langle F, F \rangle_{\mathcal{H}}$ for all $F \in \mathcal{H}^+$. But as shown above, $\langle F \circ \theta, F \circ \theta \rangle_{\mathcal{H}} = \langle F \circ \theta^2, F \rangle_{\mathcal{H}} \le \sqrt{\langle F \circ \theta^2, F \circ \theta^2 \rangle_{\mathcal{H}} \langle F, F \rangle_{\mathcal{H}}}$. By recursion, it follows that:

$$\langle F \circ \theta, F \circ \theta \rangle_{\mathcal{H}} \le \left\langle F \circ \theta^{2^n}, F \circ \theta^{2^n} \right\rangle_{\mathcal{H}}^{2^{-n}} \langle F, F \rangle_{\mathcal{H}}^{1-2^{-n}}$$

As $\left\langle F \circ \theta^{2^n}, F \circ \theta^{2^n} \right\rangle_{\mathcal{H}}^{2^{-n}} \le (\mathbb{E}(F^2(\mathcal{L}_1))^{2^{-n}}$ which converges to 1 as $n \to \infty$, the inequality follows.   □

For all $n \in \mathbb{Z}$, the symmetry $\rho^{(n)} = \theta^{-n}\rho\theta^n$ allows to define spaces $\mathcal{H}^{(n)}$ isometric to $\mathcal{H}$. These isometries can be denoted by the shift θ^n. For $n > m$, $j_{n,m} = \theta^m[\Pi^{(\theta)}]^{n-m}\theta^{-n}$ is a contraction from $\mathcal{H}^{(n)}$ into $\mathcal{H}^{(m)}$.

Chapter 10
The Case of General Symmetric Markov Processes

10.1 Overview

We now explain briefly how some of the above results can be extended to symmetric Markov processes on continuous spaces. The construction of the loop measure as well as a lot of computations can be performed quite generally, using Markov processes or Dirichlet space theory (Cf. for example [10]). It works as soon as the bridge or excursion measures $\mathbb{P}_t^{x,y}$ can be properly defined. The semigroup should have a density with respect to the duality measure given by a locally integrable kernel $p_t(x, y)$. This is very often the case in examples of interest, especially in finite dimensional spaces.

The main issue is to determine wether the results which have been developed in the previous chapters still hold, and precisely in what sense.

10.1.1 Loop Hitting Distributions

An interesting result is formula (4.9), and its reformulation in Proposition 27.

The expression on the lefthand side is well defined but the determinants appearing in (4.9) are not. In the example of Brownian motion killed at the exit of a bounded domain, Weyl asymptotics show that the divergences appearing on the righthand side of (4.9) may cancel. And in fact, the righthand side in (27) can be well defined in terms of the densities of the hitting distributions of F_1 and F_2 with respect to their capacitary measures, which allow to take the trace. A direct proof, using Brownian motion and classical potential theory, should be easy to provide, along the lines of the solution of Exercise 31.

10.1.2 Determinantal Processes

Another result of interest involves the point process defined by the points connected to the root of a random spanning tree. In the case of an interval of $\mathbb{Z}$, we get a process with independent spacings. For one dimensional diffusions, this point process with independent spacings has clearly an analogue which is the determinantal process with independent spacings (See [49]) defined by the kernel $\sqrt{k(x)}G(x,y)\sqrt{k(y)}$ (k beeing the killing rate and G the Green function). For one dimensional Brownian motion killed at a constant rate, we recover Macchi point process (Cf. [30]).

It suggests that this process (together with the loop ensemble $\mathcal{L}_1$) can be constructed by various versions of Wilson algorithm adapted to the real line. A similar result holds on $\mathbb{Z}$ or $\mathbb{N}$, where the natural ordering can be used to construct the spanning tree by Wilson algorithm, starting at 0.

For constant killing rate, $\sqrt{k(x)}G(x,y)\sqrt{k(y)}$ can be expressed as $\rho\exp(-\left|x-y\right|/a)$, with $a, \rho > 0$ and $2\rho a < 1$, the law of the spacings has therefore a density proportional to $e^{-\frac{x}{a}}\sinh(\sqrt{1-2\rho a}\frac{x}{a})$ (Cf. [30]), which appears to be the convolution of two exponential distributions of parameters $\frac{1}{a}(\sqrt{1-2\rho a}+1)$ and $\frac{1}{a}(-\sqrt{1-2\rho a}+1)$. A similar result holds on $\mathbb{Z}$ with geometric distributions. The spanning forest obtained by removing the cemetery point is composed of trees made of pair of intervals joining at points directly connected to the cemetery, whose length are independent with laws given by these (different!) exponential distributions. The separating points between these trees form a determinantal process intertwinned with the previous one (the roots directly connected to the cemetery point), with the same distribution. There are two equally probable intertwinning configurations on $\mathbb{R}$, and only one in $\mathbb{R}^+$ or $\mathbb{R}^-$.

10.1.3 Occupation Field and Continuous Branching

Let us consider more closely the occupation field $\widehat{l}$. The extension is rather straightforward when points are not polar. We can start with a Dirichlet space of continuous functions and a measure m such that there is a mass gap. Let P_t denote the associated Feller semigroup. Then the Green function $G(x,y)$ is well defined as the mutual energy of the Dirac measures δ_x and δ_y which have finite energy. It is the covariance function of a Gaussian free field $\phi(x)$, and the field $\frac{1}{2}\phi(x)^2$ will have the same distribution as the field $\widehat{\mathcal{L}}_{\frac{1}{2}}^x$ of local times of the Poisson process of random loops whose intensity is given by the loop measure defined by the semigroup P_t. This will applies to examples related to one-dimensional Brownian motion (or to Markov chains on countable spaces).

Remark 27. When we consider Brownian motion on the half line, the associated occupation field $\widehat{\mathcal{L}}_\alpha$ is a continuous branching process with immigration, as in the simple random walk case considered above.

10.1.4 Generalized Fields and Renormalization

When points are polar, one needs to be more careful. We will consider only the case of the two and three dimensional Brownian motion in a bounded domain D killed at the boundary, i.e. associated with the classical energy with Dirichlet boundary condition. The Green function does not induce a trace class operator but it is still Hilbert–Schmidt which allows us to define renormalized determinants $\det_2$ (Cf. [47]).

If A is a symmetric Hilbert Schmidt operator, $\det_2(I + A)$ is defined as $\prod(1 + \lambda_i)e^{-\lambda_i}$ where λ_i are the eigenvalues of A.

The Gaussian field (called free field) whose covariance function is the Green function is now a generalized field: Generalized fields are not defined pointwise but have to be smeared by a compactly supported continuous test function f. Still $\phi(f)$ is often denoted $\int \phi(x)f(x)dx$.

The Wick powers : ϕ^n : of the free field can be defined as generalized fields by approximation as soon as the $2n$-th power of the Green function, $G(x, y)^{2n}$ is locally integrable (Cf. [48]). This is the case for all n for the two dimensional Brownian motion killed at the exit of an open set, as the Green function has only a logarithmic singularity on the diagonal, and for $n = 2$ in dimension three as the singularity is of the order of $\frac{1}{\|x-y\|}$. More precisely, taking for example $\pi_\varepsilon^x(dy)$ to be the normalized area measure on the sphere of radius ε around x, $\phi(\pi_\varepsilon^x)$ is a Gaussian field with variance $\sigma_\varepsilon^x = \int G(z, z')\pi_\varepsilon^x(dz)\pi_\varepsilon^x(dz')$. Its Wick powers are defined with Hermite polynomials as we did previously: : $\phi(\pi_\varepsilon^x)^n$: $= (\sigma_\varepsilon^x)^{\frac{n}{2}}H_n(\frac{\phi(\pi_\varepsilon^x)}{\sqrt{\sigma_\varepsilon^x}})$. Then one can see that, for any compactly supported continuous function f, $\int f(x)$: $\phi(\pi_\varepsilon^x)^n$: dx converges in L^2 towards a limit called the n-th Wick power of the free field evaluated on f and denoted : ϕ^n : (f). Moreover, $\mathbb{E}(: \phi^n : (f) :$ $\phi^n : (h)) = \int G^{2n}(x, y)f(x)h(y)dxdy$.

In these cases, we can extend the statement of Theorem 2 to the renormalized occupation field $\widetilde{\mathcal{L}}_{\frac{1}{2}}^x$ and the Wick square : ϕ^2 : of the free field.

10.2 Isomorphism for the Renormalized Occupation Field

Let us explain this in more detail in the Brownian motion case. Let D be an open subset of $\mathbb{R}^d$ such that the Brownian motion killed at the boundary of D is transient and has a Green function. Let $p_t(x, y)$ be its transition density and $G(x,y) = \int_0^\infty p_t(x,y)dt$ the associated Green function. The loop measure μ was defined in [18] as

$$\mu = \int_D \int_0^\infty \frac{1}{t} \mathbb{P}_t^{x,x} dt$$

where $\mathbb{P}_t^{x,x}$ denotes the (non normalized) bridge measure of duration t such that if $0 \leq t_1 \leq \cdots \leq t_h \leq t$,

$$\mathbb{P}_t^{x,x}(\xi(t_1) \in dx_1, \ldots, \xi(t_h) \in dx_h) = p_{t_1}(x,x_1)p_{t_2-t_1}(x_1,x_2)\cdots p_{t-t_h}(x_h,x)dx_1 \cdots dx_h$$

(the mass of $\mathbb{P}_t^{x,x}$ is $p_t(x,x)$). Note that μ is a priori defined on based loops but it is easily seen to be shift-invariant.

For any loop l indexed by $[0\ T(l)]$, define the measure $\hat{l} = \int_0^{T(l)} \delta_{l(s)}ds$: for any Borel set A, $\hat{l}(A) = \int_0^{T(l)} 1_A(l_s)ds$.

Lemma 2. *For any non-negative function f,*

$$\mu(\langle \hat{l}, f \rangle^n) = (n-1)! \int G(x_1,x_2)f(x_2)G(x_2,x_3)f(x_3)\cdots G(x_n,x_1)f(x_1) \prod_1^n dx_i.$$

Proof. From the definition of μ and $\hat{l}$, $\mu(\langle \hat{l}, f \rangle^n)$ equals:

$$n! \int \int_{\{0<t_1<\cdots<t_n<t\}} \frac{1}{t} f(x_1)\cdots f(x_n)p_{t_1}(x,x_1)\cdots p_{t-t_n}(x_n,x) \prod dt_i dx_i dt dx$$

$$= n! \int \int_{\{0<t_1<\cdots<t_n<t\}} \frac{1}{t} f(x_1)\cdots f(x_n)p_{t_2-t_1}(x_1,x_2)\cdots p_{t_1+t-t_n}(x_n,x_1) \prod dt_i dx_i dt.$$

Performing the change of variables $v_2 = t_2 - t_1, \ldots, v_n = t_n - t_{n-1}, v_1 = t_1 + t - t_n$, and $v = t_1$, we obtain:

$$n! \int_{\{0<v<v_1,0<v_i\}} \frac{1}{v_1 + \cdots + v_n} f(x_1)\cdots f(x_n)p_{v_2}(x_1,x_2)\cdots p_{v_1}(x_n,x_1) \prod dv_i dx_i dv$$

$$= n! \int_{\{0<v_i\}} \frac{v_1}{v_1 + \cdots + v_n})p_{v_2}(x_1,x_2)\cdots p_{v_1}(x_n,x_1) \prod f(x_i)dv_i dx_i$$

$$= (n-1)! \int_{\{0<v_i\}} f(x_1)\cdots f(x_n)p_{v_2}(x_1,x_2)\ldots p_{v_1}(x_n,x_1) \prod dv_i dx_i$$

(as we get the same formula with any v_i instead of v_1)

$$= (n-1)! \int G(x_1,x_2)f(x_2)G(x_2,x_3)f(x_3)\cdots G(x_n,x_1)f(x_1) \prod_1^n dx_i.$$

$\square$

One can define in a similar way the analogous of multiple local times, and get for their integrals with respect to μ a formula analogous to the one obtained in the discrete case.

Let G denote the operator on $L^2(D, dx)$ defined by G. Let f be a non-negative continuous function with compact support in D.

Note that $\left\langle \hat{l}, f \right\rangle$ is μ-integrable only in dimension one as then, G is locally trace class. In that case, using for all x an approximation of the Dirac measure at x, local times $\hat{l}^x$ can be defined in such a way that $\left\langle \hat{l}, f \right\rangle = \int \hat{l}^x f(x) dx$.

$\left\langle \hat{l}, f \right\rangle$ is μ-square integrable in dimensions one, two and three, as G is Hilbert–Schmidt if D is bounded, since $\int \int_{D \times D} G(x, y)^2 dx dy < \infty$, and otherwise locally Hilbert–Schmidt.

N.B.: Considering distributions χ such that $\int \int (G(x, y)^2 \chi(dx) \chi(dy)$ is finite, we could see that $\left\langle \hat{l}, \chi \right\rangle$ can be defined by approximation as a square integrable variable and $\mu \left(\left\langle \hat{l}, \chi \right\rangle^2 \right) = \int (G(x, y)^2 \chi(dx) \chi(dy)$.

Let z be a complex number such that $\mathrm{Re}(z) > 0$.

Note that $e^{-z \langle \hat{l}, f \rangle} + z \left\langle \hat{l}, f \right\rangle - 1$ is bounded by $\frac{|z|^2}{2} \left\langle \hat{l}, f \right\rangle^2$ and expands as an alternating series $\sum_2^\infty \frac{z^n}{n!} \left(- \left\langle \hat{l}, f \right\rangle \right)^n$, with

$$\left| e^{-z \langle \hat{l}, f \rangle} - 1 - \sum_1^N \frac{z^n}{n!} \left(- \left\langle \hat{l}, f \right\rangle \right)^n \right| \leq \frac{\left| z \left\langle \hat{l}, f \right\rangle \right|^{N+1}}{(N+1)!}.$$

Then, for $|z|$ small enough., it follows from the above lemma that

$$\mu \left(e^{-z \langle \hat{l}, f \rangle} + z \left\langle \hat{l}, f \right\rangle - 1 \right) = \sum_2^\infty \frac{z^n}{n} Tr(-(M_{\sqrt{f}} G M_{\sqrt{f}})^n).$$

As $M_{\sqrt{f}} G M_{\sqrt{f}}$ is Hilbert–Schmidt the renormalized determinant $\det_2(I + z M_{\sqrt{f}} G M_{\sqrt{f}})$ is well defined and the second member writes $-\log(\det_2(I + z M_{\sqrt{f}} G M_{\sqrt{f}}))$.

Then the identity

$$\mu(e^{-z \langle \hat{l}, f \rangle} + z \left\langle \hat{l}, f \right\rangle - 1) = -\log(\det {}_2(I + z M_{\sqrt{f}} G M_{\sqrt{f}})).$$

extends, as both sides are analytic as locally uniform limits of analytic functions, to all complex values with positive real part.

The renormalized occupation field $\widetilde{\mathcal{L}}_\alpha$ is defined as the compensated sum of all $\hat{l}$ in $\mathcal{L}_\alpha$ (formally, $\widetilde{\mathcal{L}}_\alpha = \widehat{\mathcal{L}}_\alpha - \int \int_0^{T(l)} \delta_{l_s} ds \mu(dl)$). More precisely, we apply a standard argument used for the construction of Levy processes, setting:

$$\left\langle \widetilde{\mathcal{L}}_\alpha, f \right\rangle = \lim_{\varepsilon \to 0} \left\langle \widetilde{\mathcal{L}_{\alpha, \varepsilon}}, f \right\rangle$$

with by definition

$$\left\langle \widetilde{\mathcal{L}_{\alpha,\varepsilon}}, f \right\rangle = \sum_{\gamma \in \mathcal{L}_\alpha} \left(1_{\{T>\varepsilon\}} \int_0^T f(\gamma_s) ds - \alpha\mu(1_{\{T>\varepsilon\}} \int_0^T f(\gamma_s) ds) \right).$$

The convergence holds a.s. and in L^2, as

$$\mathbb{E}((\sum_{\gamma \in \mathcal{L}_\alpha} (1_{\{\varepsilon'>T>\varepsilon\}} \int_0^T f(\gamma_s) ds) - \alpha\mu(1_{\{\varepsilon'>T>\varepsilon\}} \int_0^T f(\gamma_s) ds))^2)$$

$$= \alpha \int (1_{\{\varepsilon'>T>\varepsilon\}} \int_0^T f(\gamma_s) ds)^2 \mu(dl)$$

and $\mathbb{E}(\left\langle \widetilde{\mathcal{L}_\alpha}, f \right\rangle^2) = Tr((M_{\sqrt{f}} G M_{\sqrt{f}})^2)$. Note that if we fix f, α can be considered as a time parameter and $\left\langle \widetilde{\mathcal{L}_{\alpha,\varepsilon}}, f \right\rangle$ are Levy processes with discrete positive jumps approximating a Levy process with positive jumps $\left\langle \widetilde{\mathcal{L}_\alpha}, f \right\rangle$. The Levy exponent $\mu(1_{\{T>\varepsilon\}}(e^{-\langle \hat{l}, f \rangle} + \left\langle \hat{l}, f \right\rangle - 1))$ of $\left\langle \widetilde{\mathcal{L}_{\alpha,\varepsilon}}, f \right\rangle)$ converges towards the Lévy exponent of $\left\langle \widetilde{\mathcal{L}_\alpha}, f \right\rangle)$ which is $\mu((e^{-\langle \hat{l}, f \rangle} + \left\langle \hat{l}, f \right\rangle - 1))$ and, from the identity $E(e^{-\langle \widetilde{\mathcal{L}_\alpha}, f \rangle}) = e^{-\alpha\mu(e^{-\langle \hat{l}, f \rangle} + \langle \hat{l}, f \rangle - 1)}$, we get the

Theorem 8. *Assume $d \le 3$. Denoting $\widetilde{\mathcal{L}_\alpha}$ the compensated sum of all $\hat{l}$ in $\mathcal{L}_\alpha$, we have*

$$\mathbb{E}(e^{-\langle \widetilde{\mathcal{L}_\alpha}, f \rangle}) = \det{}_2(I + M_{\sqrt{f}} G M_{\sqrt{f}}))^{-\alpha}.$$

Moreover $e^{-\langle \widetilde{\mathcal{L}_{\alpha,\varepsilon}}, f \rangle}$ converges a.s. and in L^1 towards $e^{-\langle \widetilde{\mathcal{L}_\alpha}, f \rangle}$.

Considering distributions of finite G^2-energy χ (i.e. such that $\int (G(x,y)^2 \chi(dx)\chi(dy) < \infty)$, we can see that $\left\langle \widetilde{\mathcal{L}_\alpha}, \chi \right\rangle$ can be defined by approximation as $\lim_{\lambda \to \infty}(\left\langle \widetilde{\mathcal{L}_\alpha}, \lambda G_\lambda \chi \right\rangle)$ and

$$\mathbb{E}(\left\langle \widetilde{\mathcal{L}_\alpha}, \chi \right\rangle^2) = \alpha \int (G(x,y))^2 \chi(dx)\chi(dy).$$

Specializing to $\alpha = \frac{k}{2}$, k being any positive integer we have:

Corollary 7. *The renormalized occupation field $\widetilde{\mathcal{L}_{\frac{k}{2}}}$ and the Wick square $\frac{1}{2} : \sum_1^k \phi_l^2 :$ have the same distribution.*

If Θ is a conformal map from D onto $\Theta(D)$, it follows from the conformal invariance of the Brownian trajectories that a similar property holds for the Brownian "loop soup" (Cf. [18]). More precisely, if $c(x) = Jacobian_x(\Theta)$

and, given a loop l, if $T^c(l)$ denotes the reparametrized loop l_{τ_s}, with $\int_0^{\tau_s} c(l_u)du = s$, the configuration $\Theta T^c(\mathcal{L}_\alpha)$ is a Brownian loop soup of intensity parameter α on $\Theta(D)$. Then we have the following:

Proposition 31. $\Theta(c\widetilde{\mathcal{L}}_\alpha)$ *is the renormalized occupation field on $\Theta(D)$.*

Proof. We have to show that the compensated sum is the same if we perform it after or before the time change. For this it is enough to check that

$$\mathbb{E}([\sum_{\gamma\in\mathcal{L}_\alpha}(1_{\{\tau_T>\eta\}}1_{\{T\leq\varepsilon\}}\int_0^T f(\gamma_s)ds - \alpha\int(1_{\{\tau_T>\eta\}}1_{\{T\leq\varepsilon\}}\int_0^T f(\gamma_s)ds)\mu(d\gamma)]^2)$$

$$= \alpha\int(1_{\{\tau_T>\eta\}}1_{\{T\leq\varepsilon\}}\int_0^T f(\gamma_s)ds)^2\mu(d\gamma)$$

and

$$\mathbb{E}([\sum_{\gamma\in\mathcal{L}_\alpha}(1_{\{T>\varepsilon\}}1_{\tau_T\leq\eta}\int_0^T f(\gamma_s)ds - \alpha\int(1_{\{T>\varepsilon\}}1_{\tau_T\leq\eta}\int_0^T f(\gamma_s)ds)\mu(d\gamma)]^2)$$

$$\alpha\int(1_{\{T>\varepsilon\}}1_{\tau_T\leq\eta}\int_0^T f(\gamma_s)ds)^2\mu(d\gamma)$$

converge to zero as ε and η go to zero. It follows from the fact that:

$$\int[1_{\{T\leq\varepsilon\}}\int_0^T f(\gamma_s)ds]^2\mu(d\gamma)$$

and

$$\int[1_{\tau_T\leq\eta}\int_0^T f(\gamma_s)ds]^2\mu(d\gamma)$$

converge to 0. The second follows easily from the first if c is bounded away from zero. We can always consider the "loop soups" in an increasing sequence of relatively compact open subsets of D to reduce the general case to that situation. □

As in the discrete case (see Corollary 3), we can compute product expectations. In dimension ≤ 3, for f_j continuous functions with compact support in D:

$$\mathbb{E}(\langle\widetilde{\mathcal{L}}_\alpha, f_1\rangle\cdots\langle\widetilde{\mathcal{L}}_\alpha, f_k\rangle) = \int Per_\alpha^0(G(x_l,x_m), 1\leq l,m\leq k)\prod f_j(x_j)dx_j. \tag{10.1}$$

10.3   Renormalized Powers

In dimension one, as in the discrete case, powers of the occupation field can be viewed as integrated self intersection local times. In dimension two, renormalized powers of the occupation field, also called *renormalized self intersections local times* can be defined, using renormalization polynomials derived from the polynomials $Q_k^{\alpha,\sigma}$ defined in Sect. 4.2. The polynomials $Q_k^{\alpha,\sigma}$ cannot be used directly as pointed out to me by Jay Rosen. See Dynkin [7,8,21,32] for such definitions and proofs of convergence in the case of paths.

Assume $d = 2$. Let $\pi_\varepsilon^x(dy)$ be the normalized arclength on the circle of radius ε around x, and set $\sigma_\varepsilon^x = \int G(y,z)\pi_\varepsilon^x(dy)\pi_\varepsilon^x(dz)$.

As the distance between x and y tends to 0, $G(x,y)$ is equivalent to $G_0(x,y) = \frac{1}{\pi}\log(\|x-y\|)$ and moreover, $G(x,y) = G_0(x,y) - H^{D^c}(x,dz)$ $G_0(z,y)$, H^{D^c} denoting the Poisson kernel on the boundary of D.

Let $G_{x,x}^{(\varepsilon)}$ (respectively $G_{y,y}^{(\varepsilon')}$, $G_{x,y}^{(\varepsilon,\varepsilon')}$, $G_{y,x}^{(\varepsilon',\varepsilon)}$) denote the operator from $L^2(\pi_\varepsilon^x)$ into $L^2(\pi_\varepsilon^x)$ (respectively $L^2(\pi_{\varepsilon'}^y)$ into $L^2(\pi_{\varepsilon'}^y)$, $L^2(\pi_\varepsilon^x)$ into $L^2(\pi_{\varepsilon'}^y)$, $L^2(\pi_{\varepsilon'}^y)$ into $L^2(\pi_\varepsilon^x)$) induced by the restriction of the Green functions to the circle pairs. Let $\iota_{x,y}^{(\varepsilon,\varepsilon')}$ be the isometry $L^2(\pi_\varepsilon^x)$ into $L^2(\pi_{\varepsilon'}^y)$ induced by the natural map between the circles.

$G_{x,x}^{(\varepsilon)}$ and $G_{y,y}^{(\varepsilon)}$ are clearly Hilbert Schmidt operators, while the products $G_{x,y}^{(\varepsilon,\varepsilon')}\iota_{y,x}^{(\varepsilon',\varepsilon)}$ and $G_{y,x}^{(\varepsilon',\varepsilon)}\iota_{x,y}^{(\varepsilon,\varepsilon')}$ are trace-class.

We define the renormalization polynomials via the following generating function:

$$\mathbf{q}_{x,\varepsilon,\alpha}(t,u) = e^{\frac{tu}{1+t\sigma_\varepsilon^x}}\det{}_2(I - \frac{t}{1+t\sigma_\varepsilon^x}G_{x,x}^{(\varepsilon)})^\alpha$$

This generating function is new to our knowledge but one should note that the generating functions of the polynomials $Q_k^{\alpha,\sigma}$ can be written $e^{\frac{tu}{1+t\sigma}}(1 - \frac{t\sigma}{1+t\sigma})^\alpha e^{\alpha\frac{t\sigma}{1+t\sigma}}$ and therefore has the same form.

Define the renormalisation polynomials $Q_k^{x,\varepsilon,\alpha}$ by:

$$\sum t^k Q_k^{x,\varepsilon,\alpha}(u) = \mathbf{q}_{x,\varepsilon,\alpha}(t,u)$$

The coefficients of $Q_k^{x,\varepsilon,\alpha}$ involve products of terms of the form $Tr([G_{x,x}^{(\varepsilon)}]^m) = \int G(y_1,y_2)G(y_2,y_3)\cdots G(y_m,y_1)\prod_1^m \pi_\varepsilon^x(dy_i)$ which are different from $(\sigma_\varepsilon^x)^m$ (but both are equivalent to $\left[\frac{-\log(\varepsilon)}{\pi}\right]^m$ as $\varepsilon \to 0$).

We have the following

Theorem 9. *For any bounded continuous function f with compact support, $\int f(x)Q_k^{x,\varepsilon,\alpha}(\langle \widetilde{\mathcal{L}_\alpha}, \pi_\varepsilon^x\rangle)dx$ converges in L^2 towards a limit denoted $\langle \widetilde{\mathcal{L}_\alpha^k}, f\rangle$ and*

$$\mathbb{E}(\langle \widetilde{\mathcal{L}_\alpha^k}, f\rangle\langle \widetilde{\mathcal{L}_\alpha^l}, h\rangle) = \delta_{l,k}\frac{\alpha(\alpha+1)\cdots(\alpha+k-1)}{k!}\int G^{2k}(x,y)f(x)h(y)dxdy.$$

Proof. The idea of the proof can be understood by trying to prove that

$$\mathbb{E}((\int f(x) Q_k^{x,\varepsilon,\alpha}(\langle \widetilde{\mathcal{L}_\alpha}, \pi_\varepsilon^x \rangle) dx)^2)$$

remains bounded as ε decreases to zero. One should expand this expression in terms of sums of integrals of product of Green functions and check that cancellations analogous to the combinatorial identities (4.7) imply the cancelation of the logarithmic divergences.

These cancellations become apparent if we compute

$$(1) = \mathbb{E}(\mathbf{q}_{x,\varepsilon,\alpha}(t, \langle \widetilde{\mathcal{L}_\alpha}, \pi_\varepsilon^x \rangle) \mathbf{q}_{y,\varepsilon',\alpha}(s, \langle \widetilde{\mathcal{L}_\alpha}, \pi_{\varepsilon'}^y \rangle))$$

which is well defined for s and t small enough. As the measures π_ε^x and $\pi_{\varepsilon'}^y$ are mutually singular $L^2(\pi_\varepsilon^x + \pi_{\varepsilon'}^y)$ is the direct sum of $L^2(\pi_\varepsilon^x)$ and $L^2(\pi_{\varepsilon'}^y)$, and any operator on $L^2(\pi_\varepsilon^x + \pi_{\varepsilon'}^y)$ can be written as a matrix $\begin{pmatrix} A & B \\ C & D \end{pmatrix}$ where A (respectively D, B, C) is an operator from $L^2(\pi_\varepsilon^x)$ into $L^2(\pi_\varepsilon^x)$ (respectively $L^2(\pi_{\varepsilon'}^y)$ into $L^2(\pi_{\varepsilon'}^y)$, $L^2(\pi_\varepsilon^x)$ into $L^2(\pi_{\varepsilon'}^y)$, $L^2(\pi_{\varepsilon'}^y)$ into $L^2(\pi_\varepsilon^x)$).

Theorem 8 can be proved in the same way for the Brownian motion time changed by the inverse of the sum of the additive functionals defined by π_ε^x and $\pi_{\varepsilon'}^y$ (its Green function is the restriction of G to the union of the two circles. Alternatively, one can extend Theorem 8 to measures to get the same result). Applying this to the function equal to t (respectively s) on the circle of radius ε around x (respectively the circle of radius ε' around y) yields

$$(1) = \det {}_2(I - \frac{t}{1+t\sigma_\varepsilon^x} G_{x,x}^{(\varepsilon)})^\alpha \det {}_2(I - \frac{s}{1+s\sigma_{\varepsilon'}^y} G_{y,y}^{(\varepsilon')})^\alpha.$$

$$\left[\det {}_2 \begin{pmatrix} I - \frac{t}{1+t\sigma_\varepsilon^x} G_{x,x}^{(\varepsilon)} & -\frac{\sqrt{st}}{\sqrt{(1+t\sigma_\varepsilon^x)(1+t\sigma_{\varepsilon'}^y)}} G_{x,y}^{(\varepsilon,\varepsilon')} \\ -\frac{\sqrt{st}}{\sqrt{(1+t\sigma_\varepsilon^x)(1+t\sigma_{\varepsilon'}^y)}} G_{y,x}^{(\varepsilon',\varepsilon)} & I - \frac{s}{1+t\sigma_{\varepsilon'}^y} G_{y,y}^{(\varepsilon')} \end{pmatrix} \right]^{-\alpha}$$

$$= \left[\det {}_2 \begin{pmatrix} I - \frac{t}{1+t\sigma_\varepsilon^x} G_{x,x}^{(\varepsilon)} & -\frac{\sqrt{st}}{\sqrt{(1+t\sigma_\varepsilon^x)(1+s\sigma_{\varepsilon'}^y)}} G_{x,y}^{(\varepsilon,\varepsilon')} \\ -\frac{\sqrt{st}}{\sqrt{(1+t\sigma_\varepsilon^x)(1+s\sigma_{\varepsilon'}^y)}} G_{y,x}^{(\varepsilon',\varepsilon)} & I - \frac{s}{1+s\sigma_{\varepsilon'}^y} G_{y,y}^{(\varepsilon')} \end{pmatrix} \right]^{-\alpha}$$

$$\cdot \left[\det {}_2 \begin{pmatrix} I - \frac{t}{1+t\sigma_\varepsilon^x} G_{x,x}^{(\varepsilon)} & 0 \\ 0 & I - \frac{s}{1+s\sigma_{\varepsilon'}^y} G_{y,y}^{(\varepsilon')} \end{pmatrix} \right]^{\alpha}$$

Note that if, A and B are Hilbert–Schmidt operators, $\det_2(I + A) \det_2 (I + B) = e^{-Tr(AB)} \det_2((I + A)(I + B))$. It follows that if, A and B' are

Hilbert–Schmidt operators such that $B'' = (I + A)^{-1}(I + B') - I$ is trace class with zero trace and AB'' has also zero trace,

$$[\det{}_2(I + A)]^{-1} \det{}_2(I + B') = \det{}_2(I + B'') = \det(I + B'')$$
$$= \det((I + A)^{-1}(I + B'))$$
$$= \det((I + A)^{-\frac{1}{2}}(I + B')(I + A)^{-\frac{1}{2}}).$$

Taking now

$$A = \begin{pmatrix} -\frac{t}{1+t\sigma_\varepsilon^x} G_{x,x}^{(\varepsilon)} & 0 \\ 0 & -\frac{s}{1+t\sigma_{\varepsilon'}^y} G_{y,y}^{(\varepsilon')} \end{pmatrix}$$

and

$$B' = \begin{pmatrix} -\frac{t}{1+t\sigma_\varepsilon^x} G_{x,x}^{(\varepsilon)} & -\frac{\sqrt{st}}{\sqrt{(1+t\sigma_\varepsilon^x)(1+t\sigma_{\varepsilon'}^y)}} G_{x,y}^{(\varepsilon,\varepsilon')} \\ -\frac{\sqrt{st}}{\sqrt{(1+t\sigma_\varepsilon^x)(1+t\sigma_{\varepsilon'}^y)}} G_{y,x}^{(\varepsilon',\varepsilon)} & -\frac{s}{1+t\sigma_{\varepsilon'}^y} G_{y,y}^{(\varepsilon')} \end{pmatrix}$$

we obtain easily that $A - B'$ is trace class as

$$Tr\left(\left| \begin{pmatrix} 0 & -G_{x,y}^{(\varepsilon,\varepsilon')} \\ -G_{y,x}^{(\varepsilon',\varepsilon)} & 0 \end{pmatrix} \right| \right) = 2Tr\left(\left| G_{x,y}^{(\varepsilon,\varepsilon')} \iota_{y,x}^{(\varepsilon',\varepsilon)} \right| \right).$$

Therefore, $B'' = (I + A)^{-1}(B' - A)$ and AB'' are trace class and it is clear they have zero trace, as both are of the form $\begin{pmatrix} 0 & S \\ R & 0 \end{pmatrix}$.

Therefore, setting $V = \sqrt{st}(I + t\sigma_\varepsilon^x - tG_{x,x}^{(\varepsilon)})^{-\frac{1}{2}} G_{x,y}^{(\varepsilon,\varepsilon')}(I + s\sigma_{\varepsilon'}^y - sG_{y,y}^{(\varepsilon')})^{-\frac{1}{2}}$,

$$(1) = \det \begin{pmatrix} I & -V \\ -V^* & I \end{pmatrix}^{-\alpha}.$$

Hence,

$$(1) = \det \begin{pmatrix} I & -V \\ 0 & I - V^*V \end{pmatrix}^{-\alpha} = \det(I - V^*V)^{-\alpha}$$

$$= \det(I - st(I + s\sigma_{\varepsilon'}^y - sG_{y,y}^{(\varepsilon')})^{-1} G_{y,x}^{(\varepsilon',\varepsilon)}(I + t\sigma_\varepsilon^x - tG_{x,x}^{(\varepsilon)})^{-1} G_{x,y}^{(\varepsilon,\varepsilon')})^{-\alpha}.$$

This quantity can be expanded. Setting, for any trace class kernel $K(z, z')$ acting on $L^2(\pi_\varepsilon^y)$,

$$Per_\alpha(K^{(n)}) = \int Per_\alpha(K(z_i, z_j), 1 \leq i, j \leq n) \prod_1^n \pi_\varepsilon^y(dz_i)$$

it equals:

$$1+\sum_{1}^{\infty}\frac{1}{k!}Per_{\alpha}(([stG_{y,x}^{(\varepsilon',\varepsilon)}(I+t\sigma_{\varepsilon}^{x}-tG_{x,x}^{(\varepsilon)})^{-1}G_{x,y}^{(\varepsilon,\varepsilon')}(I+s\sigma_{\varepsilon'}^{y}-sG_{y,y}^{(\varepsilon')})^{-1}]^{(k)})=(2)$$

Identifying the coefficients of $t^k s^l$ in (1) and (2) yields the identity

$$\mathbb{E}(Q_{k}^{x,\varepsilon,\alpha}(\langle\widetilde{\mathcal{L}_{\alpha}},\pi_{\varepsilon}^{x}\rangle)Q_{l}^{x,\varepsilon',\alpha}(\langle\widetilde{\mathcal{L}_{\alpha}},\pi_{\varepsilon'}^{y}\rangle))=\delta_{l,k}\frac{1}{k!}Per_{\alpha}([G_{y,x}^{(\varepsilon',\varepsilon)}G_{x,y}^{(\varepsilon,\varepsilon')}]^{(k)})+R_{k,l}$$

where $R_{k,l}$ is the (finite) sum of the $t^k s^l$ coefficients appearing in $\sum_{1}^{\sup(k,l)}\frac{1}{k!}$ $Per_{\alpha}(([stG_{y,x}^{(\varepsilon',\varepsilon)}(I+t\sigma_{\varepsilon}^{x}I-tG_{x,x}^{(\varepsilon)})^{-1}G_{x,y}^{(\varepsilon,\varepsilon')}(I+s\sigma_{\varepsilon'}^{y}I-sG_{y,y}^{(\varepsilon')})^{-1}]^{(k)})$ different from the term $\delta_{l,k}\frac{1}{k!}Per_{\alpha}([G_{y,x}^{(\varepsilon',\varepsilon)}G_{x,y}^{(\varepsilon,\varepsilon')}]^{(k)})$

The remarkable fact is that the coefficients of $Q_{k}^{x,\varepsilon,\alpha}$ are such that this expression involves no term of the form $Tr([G_{x,x}^{(\varepsilon)}]^m)$ or $Tr([G_{y,y}^{(\varepsilon')}]^m)$. Decomposing the permutations which appear in the expression of the α-permanent into cycles, we see all the terms are products of traces of operators of the form $\int G(y_1,y_2)\cdots G(y_n,y_1)\pi_{\varepsilon_1}^{x_1}(dy_1)\cdots\pi_{\varepsilon_n}^{x_n}(dy_n)$ in which at least two x_j's are distinct. It is also clear from the expression (2) above that if we replace $G_{x,x}^{(\varepsilon)}$ and $G_{y,y}^{(\varepsilon')}$ by $\sigma_{\varepsilon}^{x}I$ and $\sigma_{\varepsilon'}^{y}I$, the expansion becomes very simple and all terms vanish except for $l=k$, the term $\frac{1}{k!}Per_{\alpha}([G_{y,x}^{(\varepsilon',\varepsilon)}G_{x,y}^{(\varepsilon,\varepsilon')}]^{(k)})$ which will be proved to converge towards $\frac{\alpha(\alpha+1)\cdots(\alpha+k-1)}{k!}G^{2k}(x,y)=\frac{G^{2k}(x,y)}{k!}\sum_{1}^{k}d(k,l)\alpha^l$ (see Remark 7 on Stirling numbers).

To prove this convergence, and also that $R_{k,l}\to 0$ as $\varepsilon,\varepsilon'\to 0$, it is therefore enough to prove the following: □

Lemma 3. *Consider for any $x_1,x_2,\ldots,x_n$, ε small enough and $\varepsilon\leq\varepsilon_1,\ldots,\varepsilon_n\leq 2\varepsilon$, with $\varepsilon_i=\varepsilon_j$ if $x_i=x_j$, an expression of the form:*

$$\Delta=\left|\prod_{i,x_{i-1}\neq x_i}G(x_{i-1},x_i)(\sigma_{\varepsilon_i}^{x_i})^{m_i}-\int G(y_1,y_2)\cdots G(y_n,y_1)\pi_{\varepsilon_1}^{x_1}(dy_1)\cdots\pi_{\varepsilon_n}^{x_n}(dy_n)\right|$$

in which we define m_i as $\sup(h,\ x_{i+h}=x_i)$ and in which at least two x_j's are distinct. Then for some positive integer N, and $C>0$, on $\cap\{\|x_{i-1}-x_i\|\geq\sqrt{\varepsilon}\}$

$$\Delta\leq C\sqrt{\varepsilon}\log(\varepsilon)^{N}$$

Proof. In the integral term, we first replace progressively $G(y_{i-1},y_i)$ by $G(x_{i-1},x_i)$ whenever $x_{i-1}\neq x_i$, using triangle, then Schwartz inequalities, to get an upper bound of the absolute value of the difference made by this substitution in terms of a sum Δ' of expressions of the form

$$(\int(G(y_1,y_2)-G(x_1,x_2))^2\pi_{\varepsilon_1}^{x_1}(dy_1)\pi_{\varepsilon_2}^{x_2}(dy_2)\int\prod G^2(y_k,y_{k+1})\prod\pi_{\varepsilon_k}^{x_k}(dy_k))^{\frac{1}{2}}$$

$$\prod_{l}G(x_l,x_{l+1}).$$

The expression obtained after these substitutions can be written

$$W = \prod_{i, x_{i-1} \neq x_i} G(x_{i-1}, x_i) \int G(y_1, y_2) \cdots G(y_{m_i-1}, y_{m_i}) \pi_{\varepsilon_i}^{x_i}(dy_1) \cdots \pi_{\varepsilon_i}^{x_i}(dy_{m_i})$$

and we see the integral terms could be replaced by $(\sigma_\varepsilon^{x_i})^{m_i}$ if G was translation invariant. But as the distance between x and y tends to 0, $G(x, y)$ is equivalent to $G_0(x, y) = \frac{1}{\pi} \log(\|x - y\|)$ and moreover,

$$G(x, y) = G_0(x, y) - H^{D^c}(x, dz) G_0(z, y).$$

As our points lie in a compact inside D, it follows that for some constant C, for $\|y_1 - x\| \leq \varepsilon$, $|\int(G(y_1, y_2) \pi_\varepsilon^x(dy_2) - \sigma_\varepsilon^x| < C\varepsilon$. Hence, the difference Δ'' between W and $\prod_{i, x_{i-1} \neq x_i} G(x_{i-1}, x_i)(\sigma_\varepsilon^{x_i})^{m_i}$ can be bounded by $\varepsilon W'$, where W' is an expression similar to W.

To get a good upper bound on Δ, using the previous observations, by repeated applications of Hölder inequality. it is enough to show that for ε small enough and $\varepsilon \leq \varepsilon_1, \varepsilon_2 \leq 2\varepsilon$, (with C and C' denoting various constants):

(1) $\int(G(y_1, y_2) - G(x_1, x_2)^2 \pi_{\varepsilon_1}^{x_1}(dy_1) \pi_{\varepsilon_2}^{x_2}(dy_2) < C(\varepsilon 1_{\{\|x_1 - x_2\| \geq \sqrt{\varepsilon}\}} + (G(x_1, x_2)^2 + \log(\varepsilon)^2) 1_{\{\|x_1 - x_2\| < \sqrt{\varepsilon}\}})$,
(2) $\int G(y_1, y_2)^k \pi_\varepsilon^x(dy_1) \pi_\varepsilon^x(dy_2) < C |\log(\varepsilon)|^k$ and more generally
(3) $\int G(y_1, y_2)^k \pi_{\varepsilon_1}^{x_1}(dy_1) \pi_{\varepsilon_2}^{x_2}(dy_2) < C |\log(\varepsilon)|^k$.

As the main contributions come from the singularities of G, they follow from the following simple inequalities:

(1')

$$\int \left| \log(\varepsilon^2 + 2R\varepsilon \cos(\theta) + R^2) - \log(R) \right|^2 d\theta$$

$$= \int \left| \log((\varepsilon/R)^2 + 2(\varepsilon/R) \cos(\theta) + 1) \right|^2 d\theta$$

$$< C((\varepsilon 1_{\{R \geq \sqrt{\varepsilon}\}}) + \log^2(R/\varepsilon) 1_{\{R < \sqrt{\varepsilon}\}})$$

(considering separately the cases where $\frac{\sqrt{\varepsilon}}{R}$ is large or small)

(2') $\int \left| \log(\varepsilon^2(2 + 2\cos(\theta))) \right|^k d\theta \leq C |\log(\varepsilon)|^k$
(3') $\int \left| \log((\varepsilon_1 \cos(\theta_1) + \varepsilon_2 \cos(\theta_2) + r)^2 + (\varepsilon_1 \sin(\theta_1) + \varepsilon_2 \sin(\theta_2))^2 \right|^k d\theta_1 d\theta_2 \leq C(|\log(\varepsilon)|)^k$.

It can be proved by observing that for $r \leq \varepsilon_1 + \varepsilon_2$, we have near the line of singularities (i.e. the values $\theta_1(r)$ and $\theta_2(r)$ for which the expression under the log vanishes) to evaluate an integral which can be bounded (after a change of variable) by an integral of the form $C \int_0^1 (-\log(\varepsilon u))^k du \le C'(-\log(\varepsilon))^k$ for ε small enough.

To finish the proof of the theorem, let us note that by the lemma above, and the estimates in its proof, for $\varepsilon \le \varepsilon_1, \varepsilon_2 \le 2\varepsilon$, we have, for some integer $N^{l,k}$

$$\left| \mathbb{E}\left(Q_k^{x,\varepsilon_1,\alpha}(\langle \widetilde{\mathcal{L}_\alpha}, \pi_{\varepsilon_1}^x \rangle) Q_l^{y,\varepsilon_2,\alpha}(\langle \widetilde{\mathcal{L}_\alpha}, \pi_{\varepsilon_2}^y \rangle) \right) - \delta_{l,k} G(x,y)^{2k} \frac{\alpha(\alpha+1)\cdots(\alpha+k-1)}{k!} \right|$$

$$\le C \log(\varepsilon)^{N_{l,k}} (\sqrt{\varepsilon} + G(x,y)^{l+k} 1_{\{\|x-y\|<\sqrt{\varepsilon}\}}. \quad (10.2)$$

The bound (10.2) is uniform in (x,y) only away from the diagonal as $G(x,y)$ can be arbitrarily large but we conclude from it that for any bounded integrable f and h,

$$\left| \int (\mathbb{E}(Q_k^{x,\varepsilon_1,\alpha}(\langle \widetilde{\mathcal{L}_\alpha}, \pi_{\varepsilon_1}^x \rangle) Q_l^{y,\varepsilon_2,\alpha}(\langle \widetilde{\mathcal{L}_\alpha}, \pi_{\varepsilon_2}^y \rangle)) \right.$$

$$\left. - \delta_l^k G(x,y)^{2k} \frac{\alpha \cdots (\alpha+k-1)}{k!}) f(x) h(y) dx dy \right| \le C' \sqrt{\varepsilon} \log(\varepsilon)^{N_{l,k}}$$

(as $\int \int G(x,y)^{2k} 1_{\{\|x-y\|<\sqrt{\varepsilon}\}} dx dy$ can be bounded by $C\varepsilon^{\frac{2}{3}}$, for example).

Taking $\varepsilon_n = 2^{-n}$, it is then straightforward to check that $\int f(x) Q_k^{x,\varepsilon_1,\alpha} (\langle \widetilde{\mathcal{L}_\alpha}, \pi_{\varepsilon_n}^x \rangle) dx$ is a Cauchy sequence in L^2. The theorem follows. $\qquad \square$

Specializing to $\alpha = \frac{k}{2}$, k being any positive integer as before, it follows that Wick powers of $\sum_{j=1}^k \phi_j^2$ are associated with self intersection local times of the loops. More precisely, we have:

Proposition 32. *The renormalized self intersection local times $\widetilde{\mathcal{L}_{\frac{k}{2}}^n}$ and the Wick powers $\frac{1}{2^n n!} : (\sum_1^k \phi_l^2)^n :$ have the same joint distribution.*

Proof. The proof is just a calculation of the L^2-norm of

$$\int [\frac{1}{2^n n!} : (\sum_1^k \phi_l^2)^n : (x) - Q_n^{x,\varepsilon,\frac{k}{2}}(\frac{1}{2} : \sum_1^k \phi_l^2 : (\pi_\varepsilon^x))] f(x) dx$$

which converges to zero with ε.

The expectation of the square of this difference is the sum of two square expectations which both converge towards $\frac{k(k+2)\cdots(k+2(n-1))}{2^n n!} \int G^{2n}(x,y) f(x) f(y) dx dy$ and a middle term which converges towards twice the opposite value. The difficult term $\mathbb{E}((Q_n^{x,\varepsilon,\frac{k}{2}}(: \sum_1^k \phi_l^2 : (\pi_\varepsilon^x))] f(x) dx)^2)$ is given by the previous theorem. The two others come from simple Gaussian calculations (note that only highest degree term $\frac{u^n}{n!}$ of the polynomial $Q_n^{x,\varepsilon,\frac{k}{2}}(u)$ contributes to the expectation of the middle term) using identity (5.2).   □

In the following exercise, we compare the polynomials $Q_N^{\alpha,\sigma}$ and $Q_N^{x,\varepsilon,\alpha}$.

Exercise 42. (Continuation of Exercise 17)

(a) Show that $\mathbf{q}_{x,\varepsilon,\alpha}(t,u) = \sum_{l=0}^{\infty} \frac{t^l u^l}{l!}(1 + t\sigma_x^\varepsilon)^{-l} + \sum_{l=0}^{\infty}\sum_{m=1}^{\infty} \frac{u^l t^{l+m}}{l!m!}(1 + t\sigma_x^\varepsilon)^{-(m+l)} Per_{-\alpha}([G_{x,x}^\varepsilon]^{(m)})$

(b) Deduce that $Q_N^{x,\varepsilon,\alpha}(u) = \sum_{0 \le l \le N}\sum_{k \le N-l} A_{N,l,k} u^l (\sigma_x^\varepsilon)^{N-l}\alpha^k$ with $A_{N,l,k} = \sum_{m=k}^{N-l}(-1)^{N-l-k-m}\frac{(N-1)!}{l!m!(N-l-m)!(m+l-1)!}D_{m,k}^0$ and $D_{m,k}^0 = \sum C_{m,k}(k_j, \ 2 \le j \le n)\prod(Tr([\frac{G_{x,x}^{(\varepsilon)}}{\sigma_x^\varepsilon}]^j))^{k_j}$, for $k \ge 1$, $A_{N,l,0} = (-1)^{N-l-k}\frac{(N-1)!}{l!(N-l)!(l-1)!}$ and $A_{N,0,0} = 0$.

In particular, $Q_2^{x,\varepsilon,\alpha}(u) = \frac{1}{2}(u^2 - 2\sigma_x^\varepsilon u - \alpha Tr([G_{x,x}^{(\varepsilon)}]^2)$ and $Q_3^{x,\varepsilon,\alpha}(u) = \frac{1}{6}(u^3 - 6\sigma_x^\varepsilon u^2 + 6u(\sigma_x^\varepsilon)^2 - 3\alpha u Tr([G_{x,x}^{(\varepsilon)}]^2) + 6\alpha\sigma_x^\varepsilon Tr([G_{x,x}^{(\varepsilon)}]^2) - 2\alpha Tr([G_{x,x}^{(\varepsilon)}]^3))$

Exercise 43. Prove that $\lim_{\varepsilon\to 0}\frac{D_{m,k}^0}{d_{m,k}^0} = 1$

Exercise 44. Show that if we let α increase, $\mathbf{q}_{x,\varepsilon,\alpha}(t, \langle \widetilde{\mathcal{L}_\alpha}, \pi_\varepsilon^x\rangle)$, $Q_k^{x,\varepsilon,\alpha}(\langle\widetilde{\mathcal{L}_\alpha}, \pi_\varepsilon^x\rangle)$ and $\langle\widetilde{\mathcal{L}_\alpha^k}, f\rangle$ are martingales.

10.3.1   Final Remarks

(a) These generalized fields have two fundamental properties:

Firstly they are local fields (or more precisely local functionals of the field $\widetilde{\mathcal{L}_\alpha}$ in the sense that their values on functions supported in an open set D depend only on the trace of the loops on D.

Secondly, note we could have used a conformally covariant regularization to define $\widetilde{\mathcal{L}_\alpha^k}$, (along the same lines but with slightly different estimates), by taking π_ε^x to be the capacitary measure of the compact set $\{y, G^{x,y} \ge -\log\varepsilon\}$ and σ_ε^x its capacity. Then it appears that the action of a conformal transformation Θ on these fields is given by the k-th power of the conformal factor $c = $ Jacobian(Θ). More precisely, $\Theta(c^k\widetilde{\mathcal{L}_\alpha^k})$ is the renormalized k-th power of the occupation field in $\Theta(D)$.

(b) It should be possible to derive from the above remark and from hypercontractive type estimates the existence of exponential moments and introduce non trivial local interactions as in the constructive field theory derived from the free field (Cf. [48]).

(c) Let us also briefly consider currents. We will restrict our attention to the one and two dimensional Brownian case, X being an open subset of the line or plane. Currents can be defined by vector fields, with compact support.

Then, if we now denote by ϕ the complex valued free field (its real and imaginary parts being two independent copies of the free field), $\int_l \omega$ and $\int_X (\overline{\phi}\partial_\omega\phi - \phi\partial_\omega\overline{\phi})dx$ are well defined square integrable variables in dimension 1 (it can be checked easily by Fourier series). The distribution of the occupation field of the loop process "twisted" by the complex exponential $\exp(\sum_{l\in\mathcal{L}_\alpha}\int_l i\omega + \frac{1}{2}\widehat{l}(\|\omega\|^2))$ appears to be the same as the distribution of the field $\phi\overline{\phi}$ "twisted" by the complex exponential $\exp(\int_X (\overline{\phi}\partial_\omega\phi - \phi\partial_\omega\overline{\phi})dx)$ (Cf. [24]).

In dimension 2, logarithmic divergences occur.

(d) There is a lot of related investigations. The extension of the properties proved here in the finite framework has still to be completed, though the relation with spanning trees should follow from the remarkable results obtained on SLE processes, especially [20]. Note finally that other essential relations between SLE processes, loops and free fields appear in [5, 40, 57], and more recently in [43] and [44].

References

1. J. Bertoin, Levy processes. Cambridge (1996)
2. N. Biggs, Algebraic graph theory. Cambridge (1973)
3. N. Bourbaki, Algèbre. Chapitre III: Algèbre multilinéaire. Hermann (1948)
4. C. Dellacherie, P.A. Meyer, Probabilités et Potentiel. Chapitres XII-XVI Hermann. Paris. (1987)
5. J. Dubedat, SLE and the free field: Partition functions and couplings. J. Amer. Math. Soc. 22 995-1054 (2009)
6. E.B. Dynkin, Local times and Quantum fields. Seminar on Stochastic processes, Gainesville 1982. 69-84 Progr. Prob. Statist. 7 Birkhauser. (1984).
7. E.B. Dynkin, Polynomials of the occupation field and related randon fields. J. Funct. Anal. 58 20-52 (1984).
8. E.B. Dynkin, Self intersection gauge for random walks and for Brownian motion. Ann. Probability 16 1-57 (1988).
9. N. Eisenbaum, H. Kaspi, A characterization of the infinitely divisible squared Gaussian processes. Ann. Prob. 34 728-742 (2006).
10. M. Fukushima, Y. Oshima, M. Takeda, Dirichlet forms and Markov processes. De Gruyter. (1994)
11. K. Gawedzki, Conformal field theory. Lecture notes. I.A.S. Princeton.
12. G.A. Hunt, Markoff chains and Martin boundaries. Illinois J. Math. 4 313-340 (1960)
13. T. Kato, Perturbation theory for linear operators. Springer. (1966)
14. Kotani, M., Sunada, T., Zeta functions of finite graphs. J. Math. Sci. Univ. Tokyo 7 7-25 (2000).
15. J.F.C. Kingman, Poisson processes. Oxford (1993)
16. G. Lawler, A self avoiding random walk. Duke math. J. 47 655-693 (1980)
17. G. Lawler, Loop erased random walks. H. Kesten Festshrift: Perplexing problems in probability. Progr.Prob. 44 197-217 Birkhäuser (1999)
18. G. Lawler, W. Werner, The Brownian loop soup. PTRF 128 565-588 (2004)
19. G. Lawler, J. Trujillo Ferreis, Random walk loop soup. TAMS 359 767-787 (2007)
20. G. Lawler, O. Schramm, W. Werner, Conformal invariance of planar loop erased random walks and uniform spanning trees. Ann. Probability 32, 939-995 (2004).
21. J.F. Le Gall, Some prperties of planar Brownian motion. Ecole d'été de probabilités de St Flour XX Lecture Notes in Math. 1527, Springer (1990).
22. Y. Le Jan, Mesures associées à une forme de Dirichlet. Applications. Bull. Soc. Math. Fr. 106 61-112 (1978)
23. Y. Le Jan, On the Fock space representation of functionals of the occupation field and their renormalization. J.F.A. 80, 88-108 (1988)

Y. Le Jan, *Markov Paths, Loops and Fields*, Lecture Notes in Mathematics 2026,   115
DOI 10.1007/978-3-642-21216-1, © Springer-Verlag Berlin Heidelberg 2011

24. Y. Le Jan, Dynkin isomorphism without symmetry. Stochastic analysis in mathematical physics. ICM 2006 Satellite conference in Lisbon. 43-53 World Scientific. (2008)
25. Y. Le Jan, Dual Markovian semigroups and processes. Functional analysis in Markov processes (Katata/Kyoto, 1981), Lecture Notes in Math. 923, 47-75 Springer (1982).
26. Y. Le Jan, Temps local et superchamp. Séminaire de Probabilités XXI Lecture Notes in Maths 1247 176-190 Springer. (1987)
27. Y. Le Jan, Markov loops and renormalization. Annals of Probability 38, 1280–1319 (2010)
28. R. Lyons, Determinantal Probability Measures. Publ. Math. Inst. Hautes Etudes Sci. 98, 167-212 (2003)
29. R. Lyons, Y. Peres, Probability on trees and networks. Prepublication.
30. O. Macchi, The coincidence approach to stochastic point processes. Adv. Appl. Probability 7, 83-122 (1975)
31. P. Marchal, Loop erased Random Walks, Spanning Trees and Hamiltonian Cycles. E. Com. Prob. 5 39-50 (1999).
32. M.B. Marcus, J. Rosen, Markov Processes, Gaussian Processes, and Local Times. Cambridge studiesin advanced mathematics vol. 100 Cambridge University Press (2006)
33. M.B. Marcus, J. Rosen, Sample path properties of the local times of strongly symmetric Markov processes via Gaussian processes. Ann. Prob. 20, 1603-1684 (1992)
34. W.S. Massey, Algebraic Topology: An Introduction Springer (1967)
35. J. Neveu, Processus aléatoires gaussiens. Presses de l'Université de Montréal (1968)
36. Parry, W., Pollicott, M., Zeta functions and the periodic orbit structure of hyperbolic dynamics. Asterisque 187-188 Société Mathématique de France (1990)
37. J. Pitman, Combinatorial Stochastic Processes. 32th St Flour Summer School. Lecture Notes in Math.1875 Springer Berlin (2006)
38. C. Pittet, L. Saloff-Coste, On random walks on wreath products. Annals of Probability 30 948-977 (2002)
39. Quian Minping, Quian Min, Circulation for recurrent Markov chains. Z.F.Wahrsch. 59 205-210 (1982)
40. O. Schramm, S. Sheffield, Contour lines of the two dimensional discrete Gaussian free field. Acta Math. 202 (2009)
41. J.P. Serre, Arbres, amalgames, SL_2 Asterisque 46 Société Mathématique de France (1977)
42. J.P. Serre, Représentations linéaires des groupes finis. Paris Hermann (1971)
43. S. Sheffield, W. Werner, Conformal loop ensembles: Construction via Loop-soups. ArXiv math10062373.
44. S. Sheffield, W. Werner, Conformal loop ensembles: The Markovian Characterization. ArXiv math10062374.
45. Shirai, T., Takahashi, Y., Random point fields associated with certain Fredholm determinants I: fermion, Poisson ans boson point processes. J. Functional Analysis 205 414-463 (2003)
46. M.L. Silverstein, Symmetric Markov processes. Lecture Notes in Math. 426 Springer Berlin (1974)
47. B. Simon, Trace ideals and their applications. London Math Soc Lect. Notes 35 Cambridge (1979)
48. B. Simon, The $P(\phi_2)$ Euclidean (quantum) field theory. Princeton. (1974).
49. A. Soshnikov, Determinantal Random Point Fields. Russian Mathematical Surveys 55:5 923975 (2000)
50. H.M. Stark, A.A. Terras, Zeta functions on finite graphs and coverings. Advances in Maths 121 134-165 (1996)
51. P.F. Stebe, A Residual Property of Certain Groups. Proc. American Math. Soc. 26 37-42 (1970)

52. K. Symanzik, Euclidean quantum field theory. Scuola intenazionale di Fisica "Enrico Fermi". XLV Corso. 152-223 Academic Press. (1969)
53. A.S. Sznitman, Vacant set of random interlacements and percolation. Annals of Maths 171, 2039–2087 (2010)
54. D. Vere Jones, A generalization of permanents and determinants. Linear Algebra and Appl. 111 (1988)
55. D. Vere Jones, Alpha permanents and their applications. New Zeland J. Math. 26 125-149 (1997)
56. M. Weil, Quasi processus. Séminaire de Probabilités IV, Lecture Notes in Math. 124, Springer, Berlin 217-239 (1970).
57. W. Werner, The conformally invariant measure on self-avoiding loops. J. American Math Soc. 21 137-169 (2008).

Index

Y. Le Jan, *Markov Paths, Loops and Fields*, Lecture Notes in Mathematics 2026,   119
DOI 10.1007/978-3-642-21216-1, © Springer-Verlag Berlin Heidelberg 2011

Programme of the school

Main lectures

Richard Kenyon   Lectures on dimers

Vladimir Koltchinskii   Oracle inequalities in empirical risk minimization and sparse recovery problems

Yves Le Jan   Markov paths, loops and fields

Short lectures

Michael Allman   Breaking the chain

Pierre Alquier   Lasso, iterative feature selection and other regression methods satisfying the "Dantzig constraint"

Jürgen Angst   Brownian motion and Lorentzian manifolds, the case of Robertson-Walker space-times

Witold Bednorz   Some comments on the Bernoulli conjecture

Charles Bordenave   Spectrum of large random graphs

Cédric Boutillier   The critical Ising model on isoradial graphs via dimers

Robert Cope   Modelling in birth-death and quasi-birth-death processes

Irmina Czarna   Two-dimensional dividend problems

Michel Émery   Geometric structure of Azéma martingales

Christophe Gomez   Time reversal of waves in random waveguides: a super resolution effect

Nastasiya Grinberg   Semimartingale decomposition of convex functions of continuous semimartingales

François d'Hautefeuille   Entropy and finance

Erwan Hillion   Ricci curvature bounds on graphs

Wilfried Huss   Internal diffusion limited aggregation and related growth models

Jean-Paul Ibrahim   Large deviations for directed percolation on a thin rectangle

Liza Jones   Infinite systems of non-colliding processes

Andreas Lagerås   General branching processes conditioned on extinction

Y. Le Jan, *Markov Paths, Loops and Fields*, Lecture Notes in Mathematics 2026,   121
DOI 10.1007/978-3-642-21216-1, © Springer-Verlag Berlin Heidelberg 2011

Benjamin Laquerrière	Conditioning of Markov processes and applications
Krzysztof Łatuszynski	"If and only if" conditions (in terms of regeneration) for $\sqrt{n}$-CLTs for ergodic Markov chains
Thierry Lévy	The Poisson process indexed by loops
Wei Liu	Spectral gap and convex concentration inequalities for birth-death processes
Gregorio Moreno	Directed polymers on a disordered hierarchical lattice
Jonathan Novak	Deforming the increasing subsequence problem
Ecaterina Sava	The Poisson boundary of lamplighter random walks
Bruno Schapira	Windings of the SRW on $\mathbb{Z}^2$ or triangular lattice and averaged Dehn function
Laurent Tournier	Random walks in Dirichlet environment
Nicolas Verzelen	Kullback oracle inequalities for covariance estimation
Vincent Vigon	LU-factorization and probability
Guillaume Voisin	Pruning a Lévy continuum random tree
Peter Windridge	Blocking and pushing interactions for discrete interlaced processes
Olivier Wintenberger	Some Bernstein's type inequalities for dependent data

List of participants

38th Probability Summer School, Saint-Flour, France
July 6–19, 2008

Lecturers

Richard KENYON   Brown University, Providence, USA
Vladimir KOLTCHINSKII   Georgia Inst. Technology, Atlanta, USA
Yves LE JAN   Université Paris-Sud, France

Participants

Michael ALLMAN   Warwick Univ., UK
Pierre ALQUIER   Univ. Paris Diderot, F
Jurgen ANGST   Univ. Strasbourg, F
Jean-Yves AUDIBERT   ENPC, Marne-la-Vallée, F
Witold BEDNORZ   Warsaw Univ., F
Charles BORDENAVE   Univ. Toulouse, F
Cédric BOUTILLIER   Univ. Pierre et Marie Curie, Paris, F
Pierre CONNAULT   Univ. Paris-Sud, Orsay, F
Robert COPE   Univ. Queensland, St Lucia, Australia
Irmina CZARNA   Univ. Wroclaw, Poland
Bassirou DIAGNE   Univ. Orléans, F
Roland DIEL   Univ. Orléans, F
Michel EMERY   Univ. Strasbourg, F
Mikael FALCONNET   Univ. Joseph Fourier, Grenoble, F
Benjamin FAVETTO   Univ. Paris Descartes, F
Jacques FRANCHI   Univ. Strasbourg, F
Laurent GOERGEN   ENS, Paris, F
Christophe GOMEZ   Univ. Paris Diderot, F
Nastasiya GRINBERG   Warwick Univ., UK
François HAUTEFEUILLE   London, UK
Erwan HILLION   Univ. Paul Sabatier, Toulouse, F
Wilfried HUSS   Graz Univ. Technology, Austria
Jean-Paul IBRAHIM   Univ. Paul Sabatier, Toulouse, F

Liza JONES	Univ. Oxford, UK
Adrien KASSEL	ENS, Paris, F
Andreas LAGERÅS	Univ. Gothenburg & Chalmers Univ. Technology, Sweden
Benjamin LAQUERRIERE	Univ. La Rochelle, F
Krzysztof LATUSZYNSKI	Warsaw School Economics, Poland
Thierry LÉVY	ENS, Paris, F
Biao LI	Univ. Blaise Pascal, Clermont-Ferrand, F
Wei LIU	Univ. Blaise Pascal, Clermont-Ferrand, F
Karim LOUINICI	Univ. Paris Diderot, F
Sybille MICAN	Univ. München, Germany
Stanislav MINSKER	Georgia Inst. Technology, Atlanta, USA
Gregorio MORENO FLORES	Univ. Paris Diderot, F
Jonathan NOVAK	Queen's Univ., Kingston, Canada
Jean PICARD	Univ. Blaise Pascal, Clermont-Ferrand, F
Erwan SAINT LOUBERT BIÉ	Univ. Blaise Pascal, Clermont-Ferrand, F
Ecaterina SAVA	Technical Univ. Graz, Austria
Catherine SAVONA	Univ. Blaise Pascal, Clermont-Ferrand, F
Bruno SCHAPIRA	Univ. Paris-Sud, Orsay, F
Eckhard SCHLEMM	Univ. Toronto, Canada & Freie Univ. Berlin, Germany
Arseni SEREGIN	Univ. Washington, USA
Laurent SERLET	Univ. Blaise Pascal, Clermont-Ferrand, F
Frédéric SIMON	Univ. Claude Bernard, Lyon, F
Laurent TOURNIER	Univ. Claude Bernard, Lyon, F
Kiamars VAFAYI	Univ. Leiden, NL
Nicolas VERZELEN	Univ. Paris-Sud, Orsay, F
Vincent VIGON	Univ. Strasbourg, F
Guillaume VOISIN	Univ. Orléans, F
Peter WINDRIDGE	Univ. Warwick, UK
Olivier WINTERBERGER	Univ. Paris 1, F
Lorenzo ZAMBOTTI	Univ. Pierre et Marie Curie, Paris, F
Jean-Claude ZAMBRINI	GFMUL, Lisbon, Portugal

LECTURE NOTES IN MATHEMATICS   🐎 Springer

Edited by J.-M. Morel, B. Teissier; P.K. Maini

Editorial Policy (for the publication of monographs)

1. Lecture Notes aim to report new developments in all areas of mathematics and their applications - quickly, informally and at a high level. Mathematical texts analysing new developments in modelling and numerical simulation are welcome.

   Monograph manuscripts should be reasonably self-contained and rounded off. Thus they may, and often will, present not only results of the author but also related work by other people. They may be based on specialised lecture courses. Furthermore, the manuscripts should provide sufficient motivation, examples and applications. This clearly distinguishes Lecture Notes from journal articles or technical reports which normally are very concise. Articles intended for a journal but too long to be accepted by most journals, usually do not have this "lecture notes" character. For similar reasons it is unusual for doctoral theses to be accepted for the Lecture Notes series, though habilitation theses may be appropriate.

2. Manuscripts should be submitted either online at www.editorialmanager.com/lnm to Springer's mathematics editorial in Heidelberg, or to one of the series editors. In general, manuscripts will be sent out to 2 external referees for evaluation. If a decision cannot yet be reached on the basis of the first 2 reports, further referees may be contacted: The author will be informed of this. A final decision to publish can be made only on the basis of the complete manuscript, however a refereeing process leading to a preliminary decision can be based on a pre-final or incomplete manuscript. The strict minimum amount of material that will be considered should include a detailed outline describing the planned contents of each chapter, a bibliography and several sample chapters.

   Authors should be aware that incomplete or insufficiently close to final manuscripts almost always result in longer refereeing times and nevertheless unclear referees' recommendations, making further refereeing of a final draft necessary.

   Authors should also be aware that parallel submission of their manuscript to another publisher while under consideration for LNM will in general lead to immediate rejection.

3. Manuscripts should in general be submitted in English. Final manuscripts should contain at least 100 pages of mathematical text and should always include

   - a table of contents;
   - an informative introduction, with adequate motivation and perhaps some historical remarks: it should be accessible to a reader not intimately familiar with the topic treated;
   - a subject index: as a rule this is genuinely helpful for the reader.

   For evaluation purposes, manuscripts may be submitted in print or electronic form (print form is still preferred by most referees), in the latter case preferably as pdf- or zipped psfiles. Lecture Notes volumes are, as a rule, printed digitally from the authors' files. To ensure best results, authors are asked to use the LaTeX2e style files available from Springer's web-server at:

   ftp://ftp.springer.de/pub/tex/latex/svmonot1/ (for monographs) and
   ftp://ftp.springer.de/pub/tex/latex/svmultt1/ (for summer schools/tutorials).

Additional technical instructions, if necessary, are available on request from lnm@springer.com.

4. Careful preparation of the manuscripts will help keep production time short besides ensuring satisfactory appearance of the finished book in print and online. After acceptance of the manuscript authors will be asked to prepare the final LaTeX source files and also the corresponding dvi-, pdf- or zipped ps-file. The LaTeX source files are essential for producing the full-text online version of the book (see http://www.springerlink. com/openurl.asp?genre=journal&issn=0075-8434 for the existing online volumes of LNM). The actual production of a Lecture Notes volume takes approximately 12 weeks.

5. Authors receive a total of 50 free copies of their volume, but no royalties. They are entitled to a discount of 33.3 % on the price of Springer books purchased for their personal use, if ordering directly from Springer.

6. Commitment to publish is made by letter of intent rather than by signing a formal contract. Springer-Verlag secures the copyright for each volume. Authors are free to reuse material contained in their LNM volumes in later publications: a brief written (or e-mail) request for formal permission is sufficient.

Addresses:
Professor J.-M. Morel, CMLA,
École Normale Supérieure de Cachan,
61 Avenue du Président Wilson, 94235 Cachan Cedex, France
E-mail: morel@cmla.ens-cachan.fr

Professor B. Teissier, Institut Mathématique de Jussieu,
UMR 7586 du CNRS, Équipe "Géométrie et Dynamique",
175 rue du Chevaleret
75013 Paris, France
E-mail: teissier@math.jussieu.fr

For the "Mathematical Biosciences Subseries" of LNM:

Professor P. K. Maini, Center for Mathematical Biology,
Mathematical Institute, 24-29 St Giles,
Oxford OX1 3LP, UK
E-mail : maini@maths.ox.ac.uk

Springer, Mathematics Editorial, Tiergartenstr. 17,
69121 Heidelberg, Germany,
Tel.: +49 (6221) 487-8259

Fax: +49 (6221) 4876-8259
E-mail: lnm@springer.com